# COMPUTER TOOLS FOR ELECTRICAL ENGINEERS

## *MATLAB & SPICE*

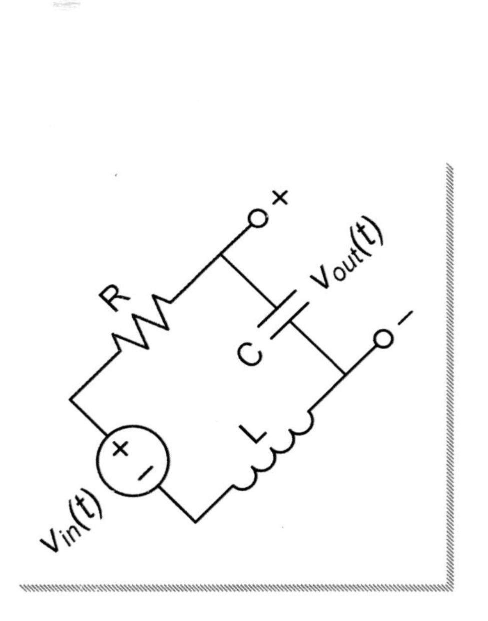


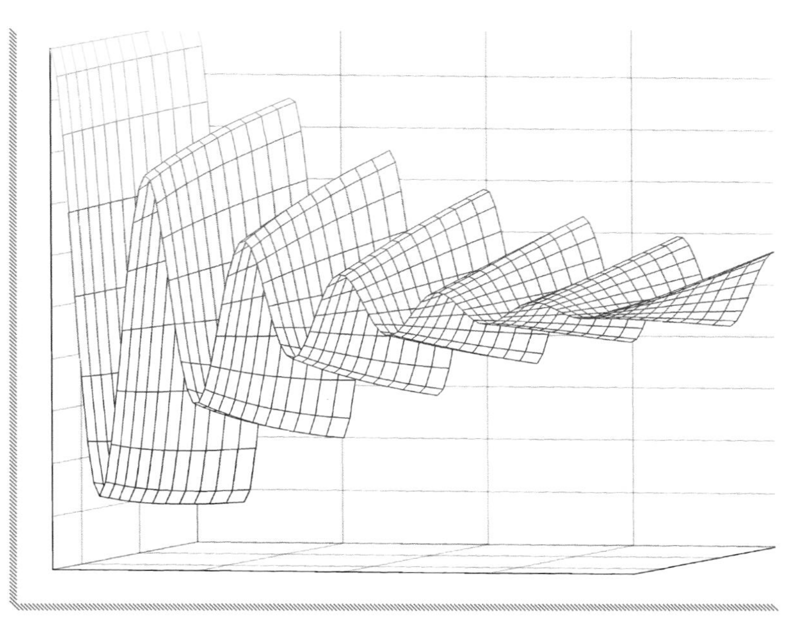




James C. Squire, P.E., Ph.D.
Julie Phillips Brown, Ph.D.

For MATLAB product information, please contact:
The MathWorks, Inc.
3 Apple Hill Drive
Natick, MA, 01760-2098
USA Tel: 508-647-7000
Fax: 508-647-7001
E-mail: info@mathworks.com
Web: *https://www.mathworks.com*
How to buy: *https://www.mathworks.com/store*
Find your local office: *https://www.mathworks.com/company/worldwide*



For LTspice product information, please contact:
Analog Devices
125 Summer Street
Boston, MA 02110-1684
USA Tel: 781-329-4700
Web: *https://www.analog.com*
How to download: *http://www.analog.com/en/design-center/design-tools-and-calculators*

ISBN: 978-1-4834-9733-4 (sc)



Lulu Publishing Services rev. date: 04/18/2019

# FOREWORD

Welcome, students, as you begin your journey to becoming electrical engineers! You probably chose this major because you wanted a challenging and rewarding college experience. Perhaps you came in search of a field of study that would suit your natural inclinations towards math and science. While some of you may have prior experiences from First Robotics, Lego League, or a high school electronics course, others may never have considered engineering as a career before a guidance counselor suggested it. Perhaps you have always known engineering was right for you, from the first time you took apart a household gizmo, or saved spare parts taken from broken toys. While most people see engineering as a way to help solve society's problems using technology, all of you will soon learn to appreciate our profession as an extended family of colleagues who share an oddball sense of humor, who have actually read their calculator manuals, and who are bothered by technical faults in science fiction movies.

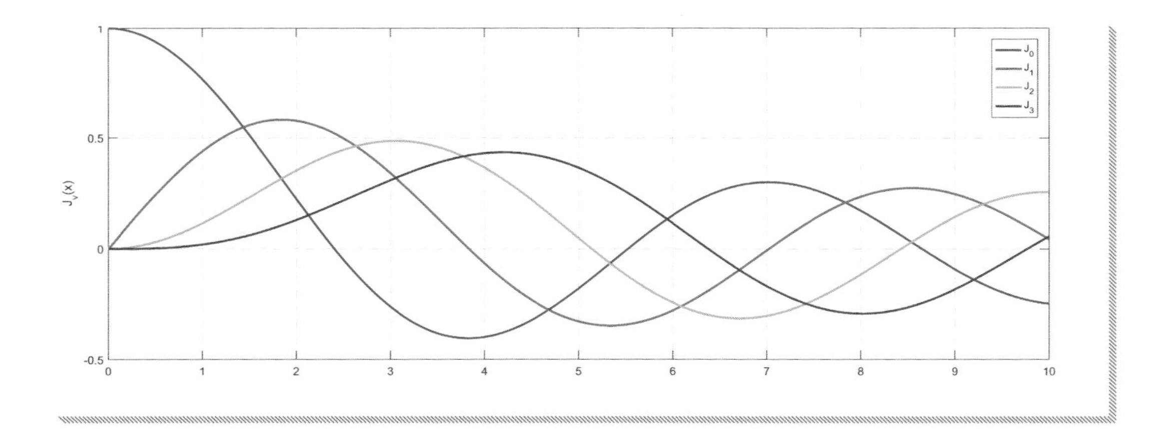

We have designed this textbook to help you learn MATLAB programming and Spice simulation skills, but also to introduce you to the culture and profession of Electrical Engineering. As you progress through the curriculum, you will find that the way you think about problems will change: your analysis will become clearer, your ability to discard confounding variables will sharpen, and your ability to construct and test models of processes will improve. This will be true whether the problem is how to build an improved voltage amplifier, how to increase the morale of the junior engineers in your unit, or how to grow a company in the face of fierce international competition. EE graduates perform well in engineering, leadership, and business roles upon graduation, and many go on to diverse careers in law, medicine, and the military. Indeed, *Forbes Magazine* notes that engineering and computer science majors were found to have higher post-graduation salaries than any other major[1]. These advantages do not attenuate over time; engineering has long been ranked as the most common undergraduate degree among Fortune 500 CEOs[2], more than business and economics majors.

While a degree in engineering confers many benefits upon graduation, most of all, we hope to convey that we become engineers not for the economic benefits, nor for the opportunity to use technology to help solve societal problems (although we are grateful for both). We become engineers because we love the challenge of solving difficult problems. That is the thought that keeps both first-year students and seasoned professionals motivated into the early morning hours, and it is a large part of the success and satisfaction you will find in whatever career you ultimately choose.

Welcome to the field,

James Squire

Julie Phillips Brown

[1] "The College Majors with the Highest Starting Salaries," *Forbes Magazine*, July 2, 2015

[2] "Why Engineers Make Great CEOs," *Forbes Magazine*, May 29, 2014.

# TABLE OF CONTENTS

**Pro Tips**

# INTRODUCTION

## PURPOSE OF THIS TEXT

We wrote *Computer Tools* because we could not find an "ideal" laboratory manual from among the currently available introductions to MATLAB and introductions to Spice tailored for students of Electrical Engineering (EE). Many such texts read like user manuals (e.g., "this key does this, this command that"), without much attention to connecting to the underlying principles of Electrical Engineering. Other texts were abstract, focusing more on the heart of procedural programming, but likely too advanced for a first-year engineering audience. A few of the texts currently available provide excellent introductions to general engineering principles, but they do not focus on the unique concerns of specific majors.

Unlike other manuals, *Computer Tools* introduces MATLAB, Spice, and general programming and analysis concepts within an Electrical Engineering context, and it tailors its content to the specific needs of a single-semester EE course. Thus *Computer Tools* is accessible for first-year students, whether or not they have a background in programming or circuit analysis. Whereas this textbook omits concepts that would be critically important to a Computer Science student, like data structures, it includes specialized mathematics applications, like complex phasor analysis, which are central to frequency-domain analysis in Electrical Engineering. *Computer Tools* also provides an overview of computationally-intensive methods used in Circuit Analysis, Electronics, Signals & Systems, Controls, and Digital Signal Processing. It touches on some ideas from the digital courses as well, with the exception of hardware description language (HDL) modeling of general digital systems, which deserves an entire course of its own. In short, this textbook offers both a general introduction to the analog aspects of electrical engineering and a specific overview of the programming packages of MATLAB and Spice.

## SOFTWARE FOR ELECTRICAL ENGINEERS

There are a number of software packages commonly used in Electrical and Computer Engineering, including MATLAB, Spice, C, C++, VHDL, Python, Java, C#, JavaScript, Multisim, and Verilog. You will learn about the first two in this text, and many more as you progress through the curriculum.

- **MATLAB**: MATLAB is the *lingua franca* of Electrical Engineering, Aeronautics, and Bioengineering, and it is becoming increasingly common in Mechanical Engineering as well. MATLAB is an advanced calculator, a graphing package with vastly more customizable results than Excel, and a general programming language. MATLAB works primarily with numbers; although it has some symbolic capabilities, it is not intended to be an algebraic solver, like Wolfram Alpha or Mathematica.
- **Spice:** Spice is a circuit simulator capable of simulating almost any analog or digital circuit. It lets the user build the circuit graphically, and then probe it to read voltages and currents. There are many variants of Spice; in this course, we will use a freeware version called LTSpice, published by Analog Devices.
- **C:** The oldest of the programming languages still widely in use, C's simple syntax is well-suited for use in embedded systems (miniaturized computers-on-a-chip). It is very closely related to the language used to program Arduino microcontrollers. Despite its ubiquity in miniaturized embedded systems, you won't find it used in full-sized PC programs because it cannot make complex applications as ably as modern programming languages that support object-oriented behavior.

- **C++:** This language (pronounced "Cee-plus-plus") is one of the most common languages used to develop programs that run on personal computers and the web. Unlike MATLAB and Python, it can run directly from the operating system as a standalone program (MATLAB and Python programs must run from within a MATLAB or Python shell). C++ is a superset of the C programming language that adds object-oriented behavior.
- **VHDL:** VHDL is a specialized language used to program a specific type of integrated circuit called a Field Programmable Gate Array (FPGA). These chips don't execute code in a typical fashion; rather, they run many thousands of processes at the same time, and they require a specialized language to do so. VHDL is slightly more common among defense contractors than its major competitor, Verilog.
- **Python:** This relatively new language is similar to an open-source version of MATLAB. Python has become increasingly common in the hobbyist world, since it does not require an expensive MATLAB license, but it is not yet as common as MATLAB in industry and academia.
- **Java:** Java is a popular multi-platform programming language. Like C++, it is object-oriented and creates stand-alone programs, and many feel it has a cleaner syntax than C++. Java is very common, but its overall use is slowly declining.
- **JavaScript:** Despite its name, JavaScript has little in common with Java. This language has grown in popularity since it first became an embedded part of HTML (which is one of the main reasons that the use of Java has declined), and it is now most commonly used within the html of web pages.
- **C#:** C# is a modern, object-oriented programming language, similar in syntax to C++, but also cleaner, like Java. It has become more popular than Java for creating Windows-based programs that run on PCs and the web, but it is tied to the Microsoft operating system.

- **Multisim:** Multisim is another popular circuit simulator, similar to Spice. It is currently more popular among hobbyists, but it is steadily gaining traction among industry professionals.
- **Verilog:** Like VHDL, Verilog is another language used to program FPGAs. Verilog is more common among non-defense contractors in the United States, and the language looks more like C.
- **MathCAD:** MathCAD is a hybrid of a spreadsheet and a higher-mathematics package, like MATLAB. It is faster to learn, but it also has less programming power. MathCAD is commonly used in Civil Engineering.
- **AutoCAD, Solidworks, and Inventor:** These are solid-modeling programs designed to create virtual models of two or three-dimensional structures. AutoCAD is very commonly used by civil engineers, while mechanical engineers generally prefer SolidWorks or Inventor. Solid modeling skills are usually not expected among electrical engineering students, but they are useful to have.

### *Why MATLAB and Spice?*

While it would be impressive to have students finish their first semester with a working knowledge of all of the above programs, we concentrate on MATLAB and Spice in this course because:

1. There is not enough time to cover all the packages and languages used by electrical engineers
2. Many of the above programs will not make sense until you take more advanced engineering courses. VHDL and Verilog, for example, first require an understanding of state machines and combinatorial logic, topics taught in Digital Logic courses.
3. You will likely find MATLAB and Spice helpful in future courses, such as Circuits, Digital Signal Processing, and Electronics.

## FORMATTING CONVENTIONS

Each chapter in *Computer Tools* has two types of problems, "Practice Problems" and "Lab Problems." Practice Problems will check your general comprehension of specific material from the preceding pages. At the end of each chapter, Lab Problems present more difficult problems that will challenge you to synthesize and implement the material you have learned throughout the previous chapter.

### *BOXED ITEMS*

Boxed items like this highlight Practice Problems, which should be completed as you read through the chapter. These relatively simple problems help significantly increase understanding and retention of the material in a way reading alone does not.

### *SHADED CALLOUTS*

Callouts like this one are useful, but not essential. They include:

**Tech Tips:** In-depth, technical discussions that may relate to more advanced subjects

**Pro Tips:** Topics relating to the profession of Electrical Engineering as a whole

**Recall:** A review of algebra skills or other materials covered earlier in the text

**Take Note:** Particularly important formulae or concepts

**Digging Deeper:** Optional, in-depth explorations of topics to enrich your introduction to Electrical Engineering

## PRACTICE PROBLEMS

1. What are the two software packages that you will learn about in this course?
2. Name two programming languages that are commonly used in Electrical Engineering and which you may learn in your career, but which are not taught in this textbook. Describe how they are used.
3. Which of the following advanced Electrical Engineering courses will build on the computationally-intensive methods you will first encounter in this course: Circuit Analysis, Semiconductors, Signals & Systems, Digital Signal Processing, Microcontrollers, C Programming, and/or Electronics?
4. Given a choice between C or C# to write a Windows program that would run on a PC, which would you use?
5. What software package would you use to write a program that runs on both Linux and Windows: Java or C#?
6. If you wanted to learn to make prototypes using a rapid-prototyping machine (also known as a solid printer), would you model them in SolidWorks or AutoCAD, assuming you were planning on sharing the work with Mechanical Engineering students?

# INTRODUCTION TO MATLAB

## OBJECTIVES

After completing this chapter, you will be able to do the following:

- Start MATLAB
- Change its working directory to your data folder
- Use MATLAB as a basic calculator
- Create and inspect variables
- Save and load variables
- Get help on syntax from within MATLAB
- Create vectors in MATLAB
- Plot data
- Export commands and graphics into Microsoft Word
- Use keyboard shortcuts, like the arrow keys

## STARTING MATLAB AND THE WORKSPACE

The MATLAB program icon is shown below. MATLAB may take a minute to load, especially if the installation includes many toolboxes, or if this is the first time it has been run. MATLAB will open its standard workspace, as shown on the following page.

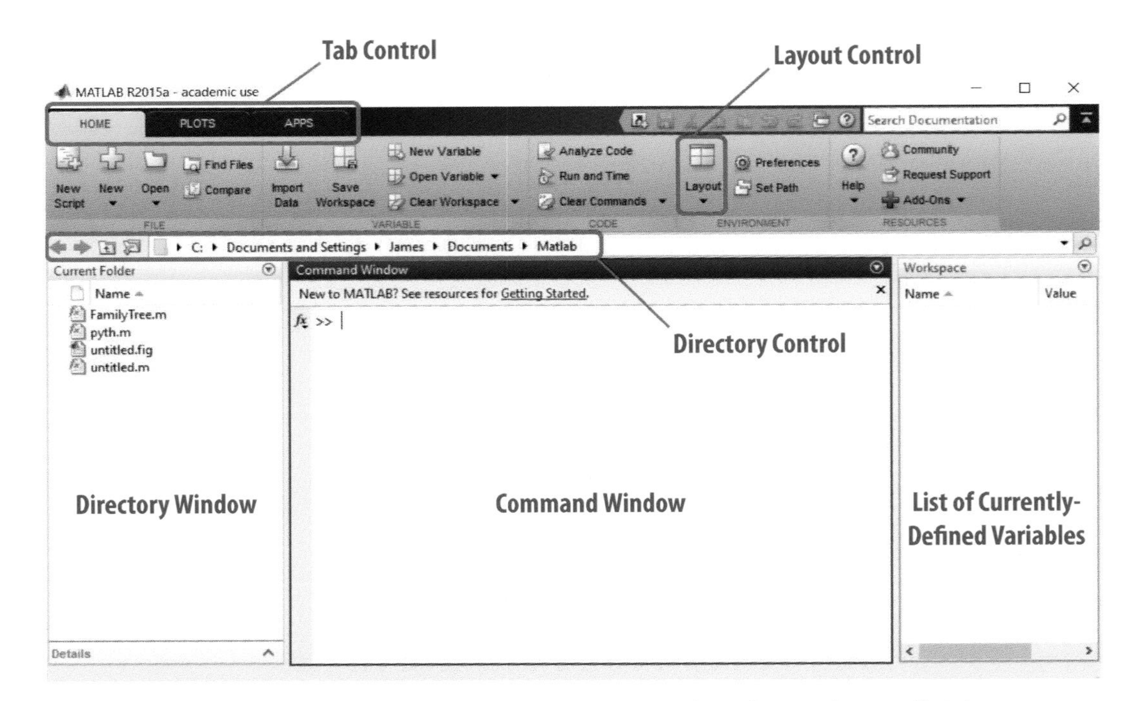


MATLAB is composed of several controls and windows. The most important window is the center "command" window, shown above, into which all MATLAB commands are entered.

If the MATLAB environment looks substantially different from the image above, set the tab control in the upper left to ***home*** and choose ***default*** from the drop-down layout control in the upper toolbar. Keep the tab control in the ***home*** position; the other positions offer various wizards (for example, to assist in plotting or filter design) that tend to be more cumbersome to use than learning the direct syntax, as is described in this text.

Next, create a data folder and make it MATLAB's current ***working directory*** (i.e. the directory in which MATLAB expects to find the user's data and programs):

1. Locate the directory control in the toolbar (top left) of the main window.
2. Use the directory control to navigate to your data directory.
3. Inside your data directory, create a new subfolder for your EE course.
4. Inside that folder, create a subdirectory called "Chapter1".
5. Note that the "Current Directory" input box in the toolbar now points to your "Chapter1" working directory.

## USING MATLAB AS A CALCULATOR

At this point, type the following commands in the command window and observe what they do. Press "Enter" after each line. The MATLAB prompt is >> and should not be typed:

```
>> 2 / 5
>> 3 + (4*9)^20
```
(MATLAB formats large numbers like $2{\cdot}10^{34}$ as 2e+34)

```
>> x = 5
>> x + 32
>> x = 2
>> y = x + 1
>> sqrt(16)
>> sqrt(-25)
```

MATLAB can be used as an expensive calculator, and it recognizes the following basic commands:

```
+
-
*
/
^
()
sqrt()
```

As in many programming languages, multiplication, division, and powers are represented by `*`, `/`, and `^`. MATLAB uses the standard order of operator precedence to evaluate expressions, so $1 + 3 \wedge 2$ is the same as $1 + (3\wedge2) = 10$, not $(1+3) \wedge 2 = 16$. When in doubt, use parentheses to be clear.

Notice that if you don't save your calculation result into a variable, MATLAB will save it into a variable called "ans," short for "answer," so you can reuse it in the next calculation, like this:

```
>> 2+3
>> ans / 10
```

returns `0.5`

MATLAB is like a high-end scientific calculator in several ways:

1. MATLAB only operates on the current line. When the value of x was changed from 5 to 2 in the previous example, unlike a spreadsheet, but like a calculator, it didn't change previous calculations involving x, but rather, only changed current and future calculations.
2. When there are variables involved (like x and y, in this example) and an equals sign, MATLAB first evaluates the statement to the right of the equals sign, and then it sets the variable to that number. Most programming languages also follow this method.

MATLAB is also unlike a calculator in other ways, which may be confusing at first:

1. MATLAB cannot perform symbolic algebra without special toolboxes. Unlike many modern calculators, which have modes that allow one to enter `x^2 - 1 = 0` and solve for `x`, MATLAB does not understand how to work with variables that are not set to a numeric value. Thus MATLAB will understand `x = 4 + 7` because `x` is set to a specific number. It will then understand `y = x + 5` and set `y` to `16`. But it will not know what to do with `2*y = 3*x`. If you want to solve for `y` in this expression, you must do it yourself using algebra, and then enter `y = 3*x/2`.
2. MATLAB does not understand implicit multiplication. `4(3+1)` will generate an error message, unlike `4*(3+1)`, which will evaluate to `16`.

### PRACTICE PROBLEMS

For every practice problem followed by a © symbol, record only the MATLAB command that solves the problem. For problems followed by a ® symbol, write the result that MATLAB returns.

1. Using MATLAB, calculate $2 \wedge 100$, and write it in scientific exponential notation (e.g. $2.4 \cdot 10^{47}$) ©®
2. Using MATLAB, calculate $2(6/13) + \frac{2.17-3.29}{2}$ ©®

## TECH TIP: CHARGE AND CURRENT

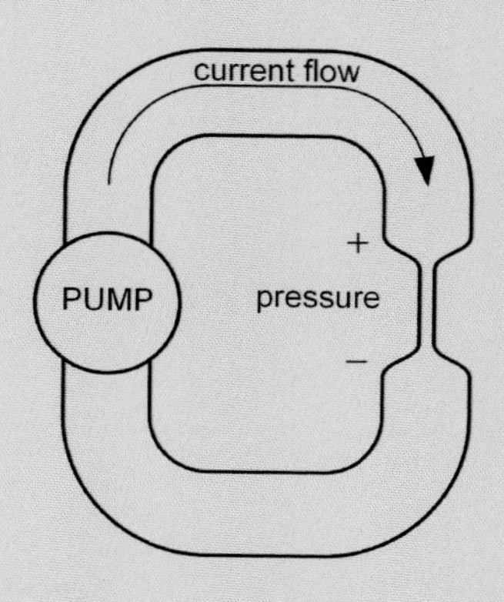


Two important quantities in ECE are current and charge. An accurate way to think of these quantities is the analogy of water being pumped through pipes around a closed circuit.

**Charge**

The amount of water (measured in gallons) is like electrical charge (measured in Coulombs, or "C"). Engineers typically use the variable "*Q*" to stand for the amount of charge, such as $Q_1 = 0.023$ C.

**Current**

The quantity of water flowing past a point (measured in gallons per second) is like electrical current (measured in Amps, or "A"). Engineers use the variable "*I*" to stand for current, such as $I_2 = 12$ A.

**Charge and Constant Current**

We can intuitively understand the relationship between a steady current flow and amount of charge that passes: if 2 gallons per second of current passed a point, in 3 seconds there would be 2 g/s * 3 s = 6 gallons of water that flowed. In the same way, if there was a constant $I = 2$ A of current that flowed in a wire for 3 seconds, there would be a total charge that passed of $Q = 2$ A * 3 s = 6 C.

For a constant current flow of $I$ Amps over a period of $t$ seconds, the general equation for total current $Q$ in Coulombs is:

$$Q = I\ t$$

Graphically, this is shown as follows:

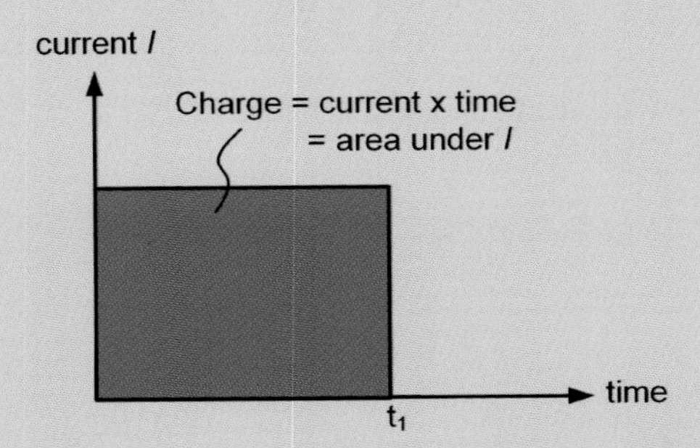


Important: Notice that instead of the formula above, the charge that flows from time 0 to time $t_1$ is also equal to the area under the current line between $t = 0$ and $t = t_1$. This is true even if the current changes with time. Think about this using the water analogy until it seems obvious.

*DIGGING DEEPER*

Those who have taken calculus will recognize that the above formula for constant current $Q = I \times t$ generalizes to:

$$Q = \int_0^{t_1} i(t)dt$$

for changing current $i(t)$. Yet the concept remains that the current $Q$ is equal to the area under the current graph.

## PRACTICE PROBLEMS

3. Using MATLAB, find the total charge that flows through a circuit with a 2.5 A current source that has been turned on for 45 seconds. (Remember units!) ©®
4. A car's headlights were left on overnight for 8 hours. If the headlights draw 0.75A, use MATLAB to calculate how much charge they drew overnight. ©®

### PRO TIP: IEEE

Electrical engineers can choose to belong to an international professional society called the Institute of Electrical and Electronic Engineers (IEEE), which is pronounced "I triple-E." This society has thousands of chapters around the world that meet monthly, typically over dinner, to talk about a topic of current engineering interest. These meetings are popular among working engineers to network, socialize, and stay current in the field. Most universities sponsor a student IEEE chapter, which may arrange inter-university robotics competitions, lead field trips, and help students find internships and ultimately, jobs. Find out more at ***ieee.org***.

## VARIABLES

MATLAB can make it easier to perform complex calculations by using variables. For example, if a design equation says the value of resistor $R_1$ can be found from the equation:

$$R_1 = \frac{1}{2\pi fC} - R_2$$

and we know $f = 2000$, $C = 20{\cdot}10^{-9}$, and $R_2 = 3000$, while one could type:

```
>> R1 = 1/(2*3.14159*2000*20e-9) - 3000
```

it is clearer and less error-prone to type:

```
>> f = 2000;
>> C = 20e-9;
>> R2 = 3000;
>> R1 = 1/(2*pi*f*C)-R2
```

Notice several things in this last example:

1. Why we're using variables: it makes the problem statement more natural. It also makes it easy to re-run the problem with different values. To see how $f = 100$ changes the calculation, for example, just change the *f* line to `f=100;` and then re-evaluate `R1 = 1/(2*pi*f*C)-R2`.
2. Pi: One can't enter Greek characters directly, but there is a built-in constant called `pi` that is defined to be $\pi$.
3. Semicolons at the end stop MATLAB from echoing the result of that line. For example, typing:

   ```
   >> f = 2000
   ```

   will cause MATLAB to parrot the following result back to the screen:

   ```
   f = 2000
   ```

   This is unnecessary when declaring variables, so suppress echoing with a semicolon, like so:

   ```
   >> f = 2000;
   ```

## NAMING AND INSPECTING VARIABLES

Variable names must begin with a letter. They may be any length, and capitalization matters. Electrical engineers tend to name resistors as R, R1, R2, R3, etc., capacitors as C, C1, C2, frequency as f, and time as t. Note that MATLAB considers R1 and r1 to be two different variables.

Variables may be composed of letters (lower and uppercase), numbers (but not as the leading character), and the underscore _ character.

Examples of valid variable names include:

```
R27
f
index
Whats_the_frequency_Kenneth
```

Invalid variable names include:

```
3R
$spent
done?
```

Once a variable is created, it stays defined until you shut down MATLAB or explicitly delete the variable. It is rare to need to delete a variable, but in case you do, it is done using the clear command, as follows:

```
>> clear('R27')
```

Every time MATLAB is run, it starts "clean," with no user-defined variables from previous sessions. To find what variables are currently defined, examine the right-most window in the MATLAB environment called the "Workspace" window. The graphic below shows R1 and f were defined in the middle "Command" window, and in the right "Workspace" window, f and R1 now appear with their current value:

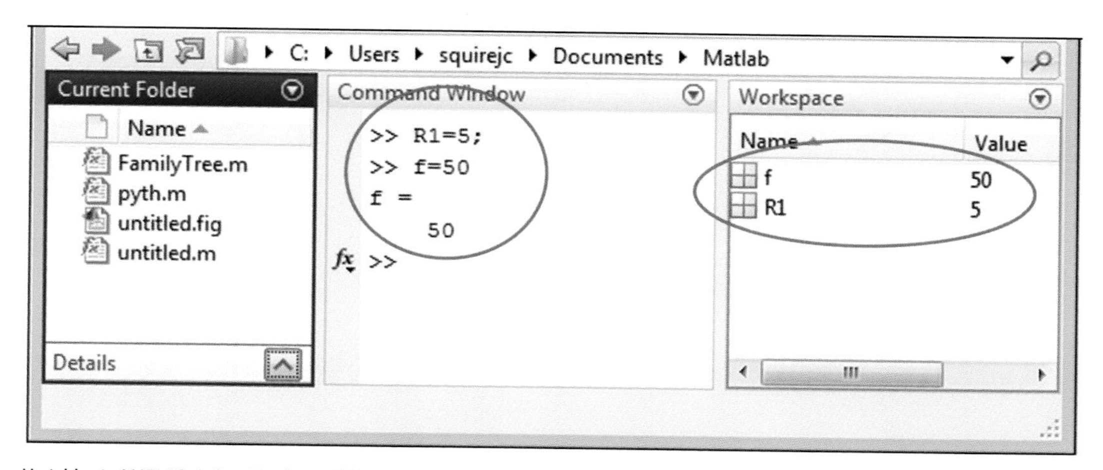


Variables in MATLAB defined in the middle command window are reflected in the workspace windows on the right.

### *PRACTICE PROBLEMS*

Which variables are valid? If they are not valid, explain why.

5. `TheAnswerIs42`
6. `SePtEmBeR21`
7. `2B_or_Not2B`
8. `R1`
9. `Current&7`

## PARENTHESES AND IMPLIED MULTIPLICATION

Unlike most calculators, MATLAB does not understand implied multiplication. Be careful to avoid entering commands like the ones shown in these examples:

```
>> 4(3+7)
```

or:

```
>> a=7;
>> 3 a
```

*Don't do this!*

will all return error messages.

Instead, multiplication must be explicitly written out with a `*`, as follows:

```
>> 4*(3+7)
```

which returns 40, or:

```
>> a=7;
>> 3*a
```

*Fixed!*

which returns 21.

## PRACTICE PROBLEMS

Use MATLAB to solve the following problems: ®

10. $4\pi\sqrt{\frac{2}{7}}$ Hint: the square root function is `sqrt()`.

11. $\frac{1}{\frac{1}{R_1}+\frac{1}{R_2}}$ if $R_1 = 6$ and $R_2 = 12$

### TECH TIP: VOLTAGE, CURRENT, CHARGE, & RESISTANCE

We already presented an analogy between water flowing in a closed-loop set of pipes and electric current flowing through a closed set of wires: the amount of water in gallons flowing past a point over a given time interval is like the charge that flows in Coulombs past a point in a wire in a time interval. The amount of water flow, measured in gallons per minute, is like the amount of charge flow, called current, which is measured in Coulombs per second, or more conveniently, in Amps. To complete the analogy, the pump that creates a pressure difference in the water to make the current flow is like a voltage source; the pressure difference between any two points is like a voltage difference; and the resistance in the piping system to the flow of water is like a resistor. In summary, these are shown with their schematic symbols in the figure on the following page.

| Quantity | Abbreviation | Units | Water Analogy |
|---|---|---|---|
| Charge | Q | Coulombs (C) | Volume of water, measured in gallons |
| Current | I | Amps (A) | Flow of water, measured in gallons per second |
| Voltage | V | Volts (V) | Pressure difference |
| Resistance | R | Ohms (Ω) | Narrow pipe causing resistance to water flow |

To create a circuit, we need the following components:

| Component | Schematic Symbol | Water Analogy |
|---|---|---|
| Wire | | Pipe |
| Voltage source | + − | Pump creating a pressure difference |
| Resistor | | Narrowed pipe resisting water flow |

A simple electrical circuit and its water analogy are shown below. Try to visualize the electrical current flow through the circuit, as it is pressurized to flow by the voltage source, and loses that pressure as it squeezes through the resistor.

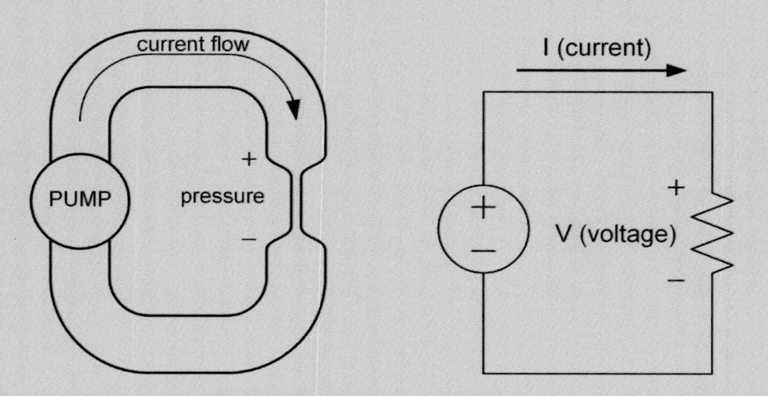


Pumped water system (left) and analogous electrical circuit (right). The pump creates a pressure difference to encourage water to flow through a restricted pipe, as the voltage source (e.g. a battery) creates a voltage difference, causing electrical current to flow through a resistor. The resistance in the pipe and circuit are both unavoidable and necessary to keep infinite current from flowing.

### PRACTICE PROBLEMS

12. Using the analogy of the pumped water system, would current increase or decrease as the resistance (of the pipe or the resistor) increases?

### TECH TIP: OHM'S LAW

About two centuries ago, Georg Ohm summarized the relationship among current, voltage, and resistance with the formula named after him. The formula is:

$$V = IR$$

where *V* is voltage, *I* is current, and *R* is resistance. If we try to understand this relationship in terms of the water analogy, it conveys two different ideas:

1. Voltage is proportional to current: as more water current is forced through a narrowed pipe, a greater pressure will develop across it; and,
2. Voltage is proportional to resistance: as the pipe narrows and its resistance increases, water flowing through it will also cause a greater pressure to develop across it.

There are two other formulations of Ohm's Law, one solved for current, shown below:

$$I = \frac{V}{R}$$

and similarly, one solved for resistance, as follows:

$$R = \frac{V}{I}$$

### *PRACTICE PROBLEMS*

13. A 68Ω resistor has 2.4A of current pumped through it. Use MATLAB to compute the amount of voltage that must be across the resistor. ©®
14. That same 68Ω resistor is now connected to a 120V voltage source. Use MATLAB to compute the amount of current that flows. ©®

## SPECIAL SYMBOLS: i, j, and π

Electrical engineers use imaginary numbers very frequently, for instance $3 + j7$, where $j = \sqrt{-1}$. This is different than the way most mathematicians would write it: $3 + 7i$. Engineers use j since i is reserved to mean current, and we place the symbol in front of the number rather than behind it. MATLAB uses both i and j to mean $\sqrt{-1}$, so $3 + j7$ would be entered in MATLAB as follows:

```
>> 3+j*7
```

Another constant that electrical engineers use surprisingly often is the ratio of a circle's perimeter to its diameter, or $\pi$. In MATLAB it is called:

```
>> pi
```

## FORMATTING NUMBERS

Internally, MATLAB keeps track of all numbers up to about 16 digits of precision, but by default only displays numbers to 4 decimal places. You can always see more digits in any answer by typing:

```
>> format long
```

By contrast, typing:

```
>> format short
```

returns MATLAB to displaying the default form, rounded to four decimal places. In electrical engineering classes, we tend to keep MATLAB in short format and typically only report 3 significant digits in our answers. This is because the components we use—resistors, capacitors, and inductors—rarely have more than 3 significant digits of accuracy in their values, and the instruments we use to measure voltage and current in undergraduate laboratories are usually limited to about 3 significant digits as well.

*DIGGING DEEPER*

Curious as to why MATLAB keeps about 16 digits of internal precision, rather than exactly 16 digits? MATLAB displays numbers in base 10, but internally works with them in base 2. It stores them in exactly 52 base-2 digits. In base 10 this is log10(2^52) or 15.6536 digits.

## SCIENTIFIC NOTATION

Very large and very small numbers are more easily read in scientific notation. For example, a typical capacitor used in our field is a $6.8 \cdot 10^{-12}$ Farad capacitor. It would be much harder to pick one out from the parts cabinet if its label read 0.0000000000068. In MATLAB, one can enter values in scientific notation using this shortcut (given for the above example):

```
>> 6.8e-12
```

Notice that there is no space between the 6.8 and the e; this is not implied multiplication (which MATLAB does not do), but rather a shortcut that gives the same result as entering 6.8*10^(-12), and yet is easier to enter and read (and internally for MATLAB, faster to process). MATLAB evaluates this input as follows:

```
>> 6.8000e-12
```

since by default, it returns 4 decimal digits in its scientific notation format. Later in this book we will introduce engineering notation, which is an even more compact way to write these values than scientific notation.

## EXPONENTIALS & THEIR INVERSES: exp, ^, sqrt, log, log10

Electrical engineers frequently use the natural exponential function, $e^x$. It is so common that rather than having a defined number for *e* and entering things as powers, MATLAB has its own built-in function called `exp(x)`. For example, to compute $e^{-1}$ in MATLAB, rather than typing 2.718^(-1), one would type:

```
>> exp(-1)
```

The inverse of the natural exponent is the natural logarithm. Some mathematical textbooks call this ln, as in ln(2), but MATLAB calls it log, as in log(2). Thus the following command will equal -3.5:

```
>> log(exp(-3.5))
```

Electrical engineers often use base-10 logarithms when analyzing signals whose amplitudes vary widely. The loudness of a signal, for example, is commonly measured in decibels, which involve base-10 logarithms. To find a logarithm to the base 10, e.g. $\log_{10}(10000) = 4$, use the MATLAB command `log10()`, as follows:

```
>> log10(10000)
```

Raising numbers other than *e* to various powers is done using the ^ symbol. For example, to represent $2^{32}$ in MATLAB, type:

```
>> 2^32
```

The square root of a number can be found in MATLAB with the `sqrt()` function. For example, $\sqrt{65536}$ (another common number in EE) is found by typing:

```
>> sqrt(65536)
```

*RECALL*

Need to find a root other than a square root? Cube roots and higher are the same as taking reciprocal exponentials. For example, $\sqrt[7]{1024} = 1024^{\left(\frac{1}{7}\right)}$, or in MATLAB, 1024^(1/7). MATLAB's ^ function will thus let you find any root.

*PRACTICE PROBLEMS*

15. Use MATLAB to evaluate $e^{-5}$. ©®

16. Use MATLAB to find $20\log_{10}\left(\frac{1}{\sqrt{2}}\right)$

    Hint: the MATLAB commmand for a square root of x is `sqrt(x)`. This number, surprisingly close to an integer, turns up frequently in filter design problems. ©®

## TRIG FUNCTIONS AND THEIR INVERSES

Electrical engineers frequently use trigonometric functions, especially cosines, to represent signals. MATLAB's trig functions are `cos()`, `sin()`, and `tan()` to find the cosine, sine, and tangent of a number given in radians. For example, cosine(π/2) is found by typing the following:

```
>> cos(pi/2)
```

Inverse trig functions are given by `acos()`, `asin()`, and `atan()`, where the result is returned in radians. For example, the inverse cosine in radians of -0.5 is:

```
>> acos(-0.5)
```

Since engineers often use degrees, alternate forms of these commands are available that take their argument in degrees: `cosd()`, `sind()`, `tand()` and their inverses that return their answer in degrees: `acosd()`, `asind()`, `atand()`. For example, to find the inverse cosine of 45°, use this MATLAB command:

```
>> acosd(45)
```

MATLAB will then return 0.7071.

RECALL

There are 360° and $2\pi$ radians in a circle. The conversion factor must therefore be the ratio, $180/\pi$ or $\pi/180$. There are more degrees in any angle than radians, so to convert to degrees multiply by $180/\pi$. There are fewer radians than degrees in an angle, so to convert to radians multiply by $\pi/180$.

## CREATING VECTORS

So far MATLAB variables have been scalars, e.g. single numeric values like a = 57. MATLAB variables can also be collections of numbers called vectors. Vectors are defined in square brackets like this:

```
>> y = [12   3   -45   2.7   pi]
```

Most common mathematical functions will operate on each of the numbers within the vector, such as `exp()`, `log()`, `cos()` and the others we have discussed so far. Using the y vector defined above,

```
>> cos(y)+1
```

yields

```
ans =
   1.8439    0.0100    1.5253    0.0959         0
```

## *TECH TIP: USING A DIGITAL MULTIMETER (DMM) TO MEASURE STEADY VOLTAGE*

**Modes:** Modern DMMs may seem imposing at first sight, but you will soon be using them like a pro. To measure a voltage, first set the selector switch to DC voltage. Possible modes usually include:

- AC (varying) voltage, marked with ∿
- DC (constant) voltage, marked with ⎓
- AC current
- DC current
- Resistance

**Range:** DMMs may autorange, like the one below, or require manual adjustment. If autoranging, it will move the decimal point in the measurement to make the reading as precise as possible. Manual-ranging DMMs split each type of measurement (voltage, current, etc.) into different ranges (i.e. a 2V-20V range and a 20V-200V range). For a manual-ranging meter, select the lowest voltage greater than the one you expect to measure (e.g. to measure a 1.5V battery, if the DMM offers ranges of 0.2V, 2V, 20V, and 200V, choose the 2V range.

**Probe Jacks:** Plug the probes into the correct jacks. DMMs have one red probe and one black probe, and there are usually three or more jacks to place them. The black probe goes into the jack labeled "COM" or "Common," and the red probe goes into the jack labeled "V" or "Volts."

**Probe Tips:** The red probe goes to the more positive end, and the black tip to the more negative end. For the DMM pictured here, the battery, nominally marked 1.5V, actually reads 1.78V. If the probe tips were reversed, so that the red probe were at the negative end of the battery, the DMM would read -1.78V.

Regularly-spaced vectors can quickly be created using a colon using the format start : increment : end. It is easiest to understand with the following examples:

```
>> x = 0 : 0.25 : 2.5
```

is equivalent to

```
>> x = [0 0.25 0.5 0.75 1 1.25 1.5 1.75 2 2.25 2.5]
```

Random vectors can also be quickly created using the `rand()` function. A vector with N numbers in it can be created using `rand(1,N)`. For instance:

```
>> x = rand(1,10)
```

returns (this time, since it will return different random numbers for you, or they would not be random)

```
x =
   Columns 1 through 6
    0.1576   0.9706   0.9572   0.4854   0.8003   0.1419
   Columns 7 through 10
    0.4218   0.9157   0.7922   0.9595
```

The next chapter will more fully explore vectors and their two-dimensional versions called matrices.

## USE OF THE SEMICOLON

The semicolon ; is commonly used in MATLAB for two different purposes:

1. It permits multiple statements to be on the same line. For instance, instead of typing

   ```
   >> R1 = 2
   >> R2 = 5
   >> Rparallel = (R1*R2)/(R1+R2)
   ```

   one could save space and type

   ```
   >> R1 = 2; R2 = 5; Rparallel = (R1*R2)/(R1+R2)
   ```

2. It suppresses output. Normally if you type:

   ```
   >> R1 = 2
   ```

   The command window echoes back on two separate lines

   ```
   >> R1 =
           2
   ```

   This is unnecessary for a simple assignment and can be suppressed by ending the statement with a semicolon.

   ```
   >> R1 = 2;
   ```

   Suppressing output is very important when generating large vectors to prevent your screen from filling with data.

   ```
   >> mydata = rand(1,2000);
   ```

## PLOTTING DATA

MATLAB has an extremely comprehensive set of plotting abilities, many of which we will explore in Chapter 3. As a brief preview, try creating a sinewave using

```
>> x = 0 : 0.1 : 4*pi;
>> y = cos(x);
>> plot(x,y)
```

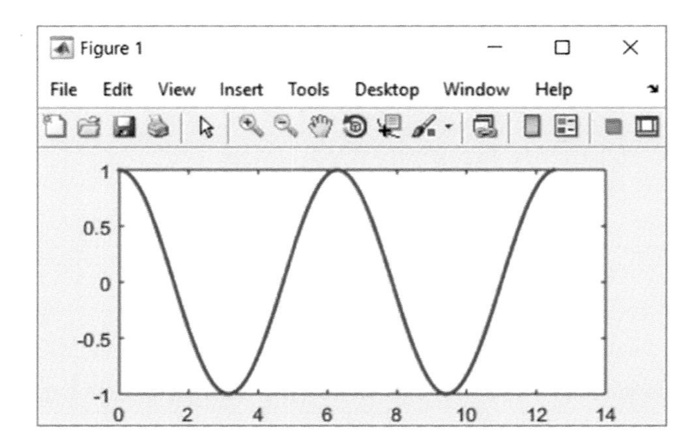


It should be clear why you want to terminate the x and y definitions with an output-suppressing semicolon; if not, enter the above commands without the terminating semicolon and observe what happens!

*PRACTICE PROBLEMS*

17. Plot a triangle with vertices at x = [0 1 .5] and y = [0 0 .7]. Hint: You may think you need just three points to define your triangle, but you'll need a fourth to close it, where the final point is a copy of the starting point. To cut and paste the figure into a Microsoft Word document, under the Edit menu in the plot choose "Copy Figure". ®

## GETTING HELP

This text is a study guide, not a MATLAB reference manual. A reference manual is an exhaustive listing of what every command does, listed alphabetically, but it does not provide instruction on how to use the system as a whole. By contrast, a study guide teaches the reader by introducing commands in small, sequential units that build upon each other. MATLAB has two different built-in reference manuals: an abbreviated manual, called "help," and a more detailed manual, called "doc. " To get quick, but abbreviated help with the syntax for a MATLAB command, type the following:

```
>> help <command>
```

For instance, you might type:

```
>> help plot
```

This will return information about the plot command in the command window:

```
plot   Linear plot.
plot(X,Y) plots vector Y versus vector X. If X or Y
is a matrix, then the vector is plotted versus the
rows or columns of the matrix, whichever line up. If
X is a scalar and Y is a vector, disconnected line
objects are created and plotted as discrete points
vertically at X.

plot(Y) plots the columns of Y versus their index.

Various line types, plot symbols and colors may be
obtained with plot(X,Y,S) where S is a character
string made from one element from any or all the
following 3 columns:

b      blue       .    point      -        solid
g      green      o    circle     :        dotted
r      red        x    x-mark     -.       dashdot
c      cyan       +    plus       --       dashed
m      magenta    *    star       (none)   no line
```

You can get more comprehensive help by typing:

```
>> doc <command>
```

For instance,

```
>> doc plot
```

will spawn a help window with detailed description of the desired command.

## SAVING AND LOADING VARIABLES

MATLAB does not remember variables between sessions; that is, when MATLAB is started, it starts with no variables defined, except for a few hidden constants like `pi`. Once variables are defined, they may be saved or loaded. Variables worth saving are typically large vectors or matrices holding a lot of data. These may be very large; for example, a vector defined to hold all the close-of-day stock prices for IBM would have over 20,000 numbers in it.

To save a variable x, which may be real or complex, scalar or vector, in a file called "x.txt" in ASCII format (which is a format that you can read and edit in a program like Notepad), enter the following:

```
>> save('x'.txt, 'x', '-ascii')
```

The filename comes first, encased in single quotes, followed by your variable in single quotes, and then the '-ascii' option to save the data in an easily-read and modified format, with each of these parts separated by a comma. Omitting the '-ascii' option will save the data as a compressed .mat file, which takes less disk space but is essentially unreadable by any program other than MATLAB. If the above syntax seems difficult to remember, don't worry. As long as you know the name and purpose of the `save()` command, you can get help by typing:

```
>> help save
```

*NOTE*

MATLAB uses single quotes for strings, not double quotes.

If you close and restart MATLAB, thus clearing the variable x from memory, you can reload it by typing:

```
>> load('x.txt')
```

This is also a simple way to import data generated by external sources into MATLAB. More complex data can be imported by right-clicking the data file in the folder window and choosing "Import Data".

## KEYBOARD SHORTCUTS

Very often you will need to repeat a sequence of keystrokes exactly or almost exactly. This has probably happened to you already, and you will find these keyboard shortcuts very convenient.

To recall previous commands, press the up arrow key ↑. Keep pressing to go further back in the command history. Once the command is visible you can edit it or run it again by pressing "Enter."

Another way to re-run a command is to look in the command history window, circled in the screen shot below. While this is easier to remember than using the up arrow key, the arrow key is much faster.

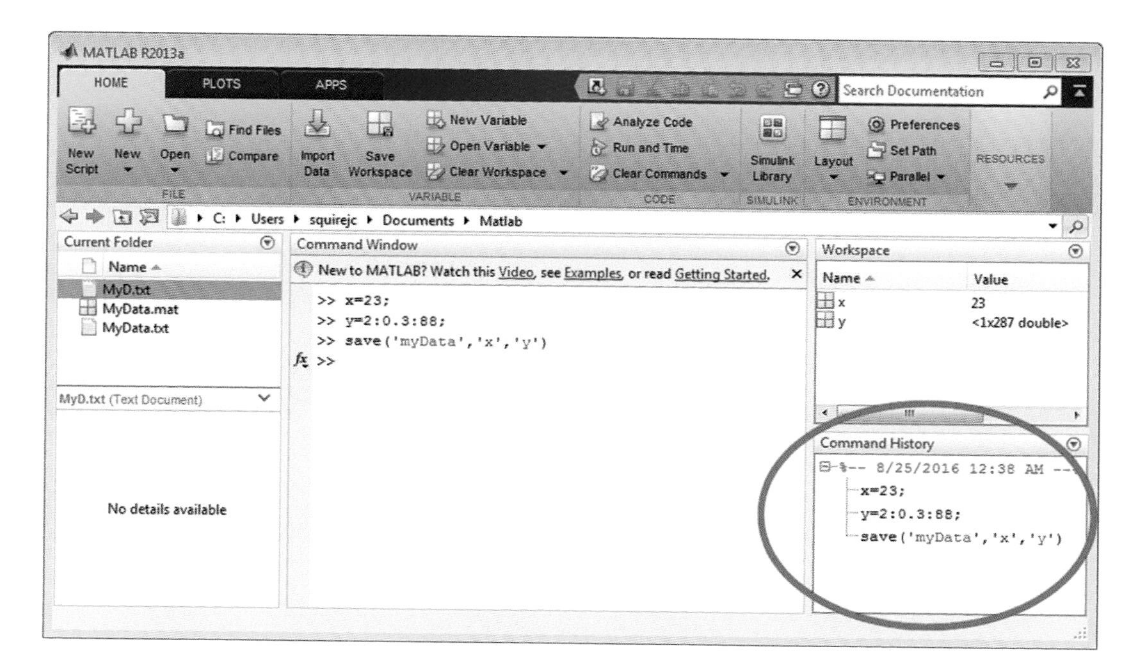

### *PRO TIP: CAREERS IN ELECTRICAL ENGINEERING*

Careers in electrical engineering can be examined by sector, title, or by subdiscipline. "Sector" refers to who hires the engineer: a company, the government, or a university. "Title" refers to the job description, which is a function of the sector. "Subdiscipline" refers to the specific electrical engineering field, such as microelectronics, power, or telecommunication.There are three primary sectors that hire electrical engineers; these are private industry, government, and academia (in order of the number of annual hires), and each has unique benefits. Salaries tend to be highest in private industry, while government positions generally offer the greatest job security, and engineers working in academic institutions enjoy significant independence in choosing research topics.

Each sector offers different types of job titles. Private industry has a great variety, including:

- Research and development: Develop prototypes of new devices and technologies;
- Production: Optimize manufacturing processes to reduce cost;
- Sales: Work as part of a team to sell a technological solution to a client;
- Project Management: Manage teams of engineers and other specialists as they solve problems too broad in scope for a single engineer; and,
- Consulting: Help other companies solve technological problems, often involving production and project management.

Government jobs often involve:

- Project Management: Oversee private industries as they perform government contract work; and,
- Testing: Determine if products meet safety and health requirements, which can be difficult to ascertain when the technology under review is novel.

Academic positions typically include:

- Research and Development: Often with a longer time-horizon to implementation than R&D positions in private industry; and,
- Teaching: Both undergraduate and graduate students.

Any employment sector can be involved with any of the main subdisciplines of Electrical Engineering. Major EE sub-disciplines include:

- Microelectronics: Fabrication of extremely small integrated circuits;
- Power: Generation and transmission of large amounts of electrical power over long distances;
- Controls: Use programmable logic or microcontrollers to precisely control motion or heat;
- Instrumentation: Design systems that measure electrical quantities or pressure, flow, and temperature;
- Telecommunications: Systems that transmit data by wire, optical fiber, or radio wave; and,
- Computer Engineering: Design of computers and associated hardware.

## COMMAND REVIEW

***Basic Math Operations***

`+ - * / ^` * means multiply. No implied multiplication

`sqrt()` square root

`log()` natural log

`log10()` log base 10

`exp()` natural exponent

***Pre-defined Constants and Notation***

`i, j` sqrt(-1), can be used for instance as 3 + j*7;

`pi` $\pi$

***Formatting Numbers***

`68e-12` type of scientific notation, here representing $68 \times 10^{-12}$

`format long` shows all significant digits (about 16)

`format short` shows about 4 significant digits of answer

***Saving and Loading Data***

`save('x.txt', 'x', '-ascii')` saves variable x in an ascii-readable file x.txt

`save('x.mat','x')` saves variable x in a compressed MATLAB format x.mat

`load('x.txt')` loads the variable in x.txt back into variable x

***Trigonometric Functions***

`cos(), sin(), tan()` cosine, sine, and tangent functions given an argument in radians

`acos(), asin(), atan()` inverse functions of the above, returning an answer in radians

`cosd(), sind(), tand()` cosine, sine, and tangent functions given an argument in degrees

`acosd(), asind(), atand()` inverse functions of the above, returning an answer in degrees

### *Vector Functions*

`[2 4 -6]` creates a vector with components 2, 4, and -6

`12:3:24` shorthand for [12 15 18 21 24]. The middle increment number may be negative; 10:-1:0 counts down from 10 to 0.

`rand(1,17)` a vector of 17 random numbers, each independently ranging from [0 1)

### *Plots*

`plot(x,y)` plots the set of lines whose horizontal positions are in vector x and whose vertical positions are in vector y.

### *Keyboard Shortcuts*

↑ (up arrow) recall last command

↑ ↑ recall second to last command, etc.

## LAB PROBLEMS

These problems require more thought than the embedded Practice Problems that were designed to check general comprehension. For each question record the MATLAB command if there is a © next to the problem, and the MATLAB result if there is a ® next to the problem. Example of a question and answer with both a ©® requirement:

| | |
|---|---|
| Question: | Use MATLAB to evaluate $4+\sqrt[6]{5}$ ©® |
| Answer: | Command: 4+(5^(1/6)) |
| | Result: 5.3077 |

1. Computers use sets of binary digits, or bits, to represent information. A bit can be 0 or 1, so a single bit can store 2 unique states. Two bits can store four states, namely 00, 01, 10, or 11. Many microprocessors store information in sets of 8 bits, called a byte. How many unique states can be represented in a byte? Hint: it is much more than 16. Most modern desktop computers store information in sets of 64 bits, called a 64-bit word. How many unique states can that represent? ©®
2. $\sqrt{3}+\sqrt[3]{4}$ Hint: read the shaded "Recall" box on page 17. ©®
3. Use MATLAB to determine what famous irrational number, used frequently in signal processing, that this continued fraction is approximating: ©®

   $$3+\frac{1}{7+\frac{1}{16}}$$

4. Using MATLAB, find the total charge that flows through a circuit with a 2.5 A current source that has been turned on for 45 seconds. (Remember to give units in your answer!) ©®
5. AA standard alkaline AA batteries can deliver 0.02A of current to power an LED light for 120 hours (surprising, but true!). How much charge circulates through the battery in this time? Remember: the current/time relationship on page 6 assumes time is measured in seconds. ©®

6. If you cut a wire and separated the ends in the schematic to the right, all current stops flowing. What is the equivalent of cutting the wire in the pipe analogy? Hint: it isn't just cutting the pipe because then water would flow everywhere.

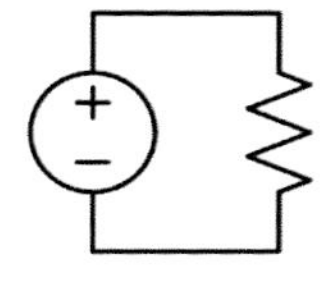


7. A 120V voltage source is connected across a resistor and 6.8 Amps of current flows. Use MATLAB to compute the value of the resistor. ©®
8. A 12Ω resistor is connected to a 14V source. Use MATLAB to compute how much charge flows through the resistor after 2 seconds. Hint: you will need two formulae. ©®
9. Engineers use some unusual numbers; *e* and *π* are both irrational numbers (about 2.71 and 3.14 respectively), and the imaginary number $j=\sqrt{-1}$ is so unusual that even its discoverer, Euler, said it was merely a mathematical curiosity of no practical use. But they combine in a surprising way that later courses will put to good use. Use MATLAB to evaluate the following: $e^{j2\pi}$ and write in the simplest possible form. Hint: it is surprisingly neither complex nor irrational. ©®
10. Use MATLAB to find the cosine of *π*/6 radians. ©®
11. Use MATLAB to evaluate the cosine of 60°. ©®
12. Create a plot with the x axis extending from 1 to 50 of 50 random numbers. Each of the 50 numbers should be between 0 and 1. Hint: you need an x vector that counts from 1 to 50, and a y vector that has the random numbers. Show the commands you used to make them (do not waste paper showing the contents of the vectors) and include the plot. ©®
13. Plot the first letter of your last name. If you are working in a lab group, choose one of your lab partners. It may be very rough; if your last name begins with "S" for example, a jagged reversed "Z" shape is fine, for instance, as long as it is recognizable. Some letters may be difficult to see because they fall on the axis boundaries; that is fine for now. You will learn in Chapter 3 how to create a margin around the plot area using the `axis()` command. ®

# MATLAB AS A CALCULATOR

2

## OBJECTIVES

After completing this chapter, you will be able to use MATLAB as an advanced calculator to perform these operations:

- Round and separate a number into its integer and fractional components
- Build and manipulate vectors
- Perform vector mathematics
- Work with complex numbers (Signals and Systems)
- Change rectangular to polar complex number formats (Circuit Analysis)
- Work with strings
- Build and manipulate matrices
- Solve large simultaneous equations (Circuit Analysis)

## ROUNDING AND SEPARATING NUMBERS

`round()` Rounds a given number to its nearest whole integer. For example, `round(3.4)` returns `3`

`floor()` Returns the integer component of a number without rounding. For example, `floor(-2.7)` returns `-2`

To find the fractional component of a number you can subtract the integer portion of the number from it. For example, `3.7 - floor(3.7)` returns `0.7`

## VECTORS

A vector is simply a list of numbers, e.g. [3, -2.4, π, $\sqrt{2}$, 7-j2]. It can hold any amount of numbers, including the special cases of 0 or 1 length. A 0-length vector is called a *null* or *empty* vector and holds nothing. A vector of length 1 is simply a number, also known as a *scalar*.

All STEM disciplines use vectors. A vector of length 4 may be a point in four-dimensional space to a mathematician, a description of a unique event in spacetime to a physicist, a set of voltages at four nodes to an electrical engineer, or a patient's four most recent blood pressure readings to a physician. Regardless of the application, it can be useful to "bundle up" the component numbers into a single list.

## CREATING VECTORS

There are several different methods for creating vectors in MATLAB:

- Defining components explicitly
- Filling with duplicate values
- Filling with linearly-spaced values
- Filling with exponentially-spaced values
- Filling with random values

### *Defining Components Explicitly*

Vectors can name each component in the list explicity by placing them inside square brackets. For example: `[3 -2.4 pi sqrt(3) 7-2*j]`

NOTE

MATLAB allows embedded constants such as pi, expressions such as sqrt(3), and complex values, like 7-j*2.

### *Filling with Duplicate Values*

Entering duplicate values like [3 3 3 3] works for just a few values, but it would be too time-consuming to use this method to create a vector of 1,000 identical numbers. Instead, use these commands:

`zeros(1,5)` creates a vector of five zeros, and is the same as typing `[0 0 0 0 0]`

`ones(1,12)` creates a vector of 12 ones, and is the same as typing `[1 1 1 1 1 1 1 1 1 1 1 1]`

NOTE

MATLAB permits multiplication of a scalar with a vector. When used in combination with `ones()`, this lets one fill a vector with duplicates of any number.

In the following examples,

`4*ones(1,5)` is the same as `[4 4 4 4 4]` and

`-pi*ones(1,2)` is the same as `[-3.14159 -3.14159]`

***Filling with Linearly-Spaced Values***

There are two ways to fill a vector with linearly-spaced values (e.g. `[1 2 3 4]`), depending on whether you know the increment between numbers or the total number of values you need. If you know the increment use the `:` operator `[start:increment:end]`. For example:

`0:0.5:3` is the same as `[0 0.5 1 1.5 2 2.5 3]`

You can decrement using a negative increment. For example:

`3:-1:0` gives the countdown equivalent to `[3 2 1 0]`

You can increment by 1 using this shortcut: `[start:end]`. For example:

`-2:2` is the same as `-2:1:2`, which yields `[-2 -1 0 1 2]`

If you do not know the increment, but know the total number of values, use `linspace`, e.g. `linspace(start, end, number_of_values)`.
For example:

`linspace(0,3,7)` returns `[0 0.5 1 1.5 2 2.5 3]`

`linspace()` can also decrement if the end value is less than the start value.
For example:

`linspace(10,0,5)` returns `[10 7.5 5 2.5 0]`

***Filling with Exponentially-Spaced Values***

ECE problems often require a list of numbers to be spaced exponentially to fill a broad range (e.g. all frequencies from 1Hz to 1MHz). MATLAB calls the command to exponentially space values `logspace`. To exponentially space 75 numbers from 100 to 100,000 (10^2 to 10^5), the command is:

`logspace(2,5,75)`

To space 5 numbers between $10^0=1$ and $10^4=10{,}000$, type:

`logspace(0,4,5)` which returns `[1 10 100 1000 10000]`

In general, to create N exponentially-spaced numbers from $10^a$ to $10^b$, type:

`logspace(a,b,N)`

NOTE

`logspace(a,b,N)` spaces the numbers from $10^a$ to $10^b$, rather than between a and b, as `linspace` does.

### *Filling with Random Values*

You have already been introduced to random number generation in Chapter 1. It is common to need random numbers in electrical engineering problems, such as when you need to simulate the effects of random variances of resistor values.

| | |
|---|---|
| `v = rand(1,10)` | creates a vector v filled with 10 random real numbers between 0 and 1. The first number, 1, directs the `rand()` command to create a vector. The second number is the number of random values to be made on the interval [0, 1). |
| `10*rand(1,20)-5` | creates a vector with 20 random numbers, each between -5 and 5. |

DIGGING DEEPER

The numbers produced by the MATLAB random generator aren't truly random—they are *pseudorandom*, since they are generated deterministically by an algorithm. This algorithm uses a *seed* number to generate its initial pseudorandom value; each successive pseudorandom number then seeds the next value generated. When MATLAB launches, it initializes its seed according to the current date and time, measured in hundredths of a second, so it always appears to generate a new sequence of random numbers.

## PRACTICE PROBLEMS

Write the MATLAB commands to generate the following vectors (do not write the output):

1. Filled with 25 values all equal to 1. ©
2. Filled with 25 values all equal to π. ©
3. Filled with 10 numbers evenly spaced from 10 to 20. ©
4. Filled with 10 random numbers between -0.5 and +0.5. ©
5. Filled with the following 3 numbers: 0, 2π, and 4. ©
6. Filled with as many numbers as it takes to span 5.125 to 17.5 in increments of 0.125. ©
7. Filled with 20 numbers, exponentially-spaced from 1 to 1,000. Hint: Does the answer look strange? For arrays with large numbers, MATLAB often writes them all as smaller numbers, with one large constant in front written in scientific notation multiplying them all. ©

### TECH TIP: MEASURING RESISTANCE

In the last chapter, you learned to measure voltages with a digital multimeter (DMM) by setting the mode to voltage, limiting the range to the smallest that did not cause an out-of-range error, and then placing the black probe in the COM jack and the red probe in the shared voltage/resistance jack. Measuring resistance uses the same approach as setting the range and uses the same probe jacks, but now the mode selector is set to resistance (abbreviated with the unit Ω).

Modern multimeters typically measure from 1 Ω up to 1 MΩ (a million Ohms) or more. The size of that span is amazing—it is like a kilometer-long ruler marked in millimeters!

The exact value of a resistor is guaranteed to be within a percentage of its marked value. This percentage is called a tolerance; common ones are 1% and 5%. Since it does not make sense to stock resistors in increments of 1% if their tolerance is only guaranteed to the nearest 5%, resistor values are available in standardized sets of values in increments approximately equal to their tolerance. For example, the first few 5% tolerance resistors values (omitting their Ω units) are: 1.0, 1.1, 1.2, 1.3, 1.5, 1.6, 1.8, 2.0, 2.2, 2.4, 2.7, 3.0, 3.3, 3.6, 3.9, 4.3, 4.7, 5.1, 5.6, 6.2, 6.8, 7.5, 8.2, 9.1, and so on in multiples of powers of 10 of these values (e.g. the next values are 10, 11, 12, 13, 15…., 100, 110, 120, 130, 150…). MATLAB could easily store this data as a single vector. Since these are close to, but not exactly 5% increments, you would simply manually enter the first 24 values (from 1.0 to 9.1) into a MATLAB vector. You could multiply that vector by the scalar 10 to find the next set of resistors from 10 to 91. In this chapter you will learn how to concatenate these two vectors to create a vector with all the resistor values from 1 to 91 Ω.

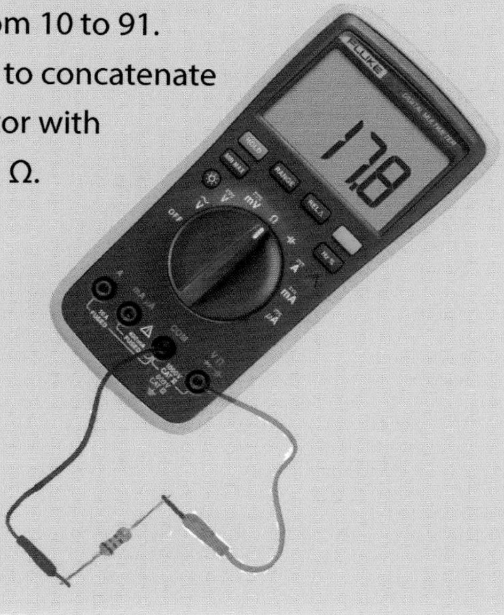

## WORKING WITH VECTORS

MATLAB offers several methods for working with vectors. The examples below assume the vector is stored in variable `v`.

### *Viewing a Vector*

To view a vector, type the variable that stores the vector. For example:

```
v
```

### *Not Viewing a Vector*

When a vector is created in MATLAB, the result is echoed to the screen by default. This can be a problem if you are creating a vector with several thousand elements. To suppress the echoing, end the command with a semicolon, e.g.:

```
v = ones(1,1000);
```

NOTE

Remember that typing a semicolon at the end will keep MATLAB from printing all 1,000 numbers when you create the vector v.

### *Vector Length*

| | |
|---|---|
| `length(v)` | Returns the number of elements in a vector.<br>For example, `length([7 6 1])` returns `3` |
| `N = length(y)` | Sets variable `N` to the number of data points present in `y` |

### *Accessing and Changing Values in Vectors*

MATLAB handles vectors differently from other programming languages in two important ways:

- It indexes vectors starting from 1, not 0. The first value in vector v is found at the first index. Most computer languages say the first value is at index 0.
- MATLAB uses parentheses ( ), not square brackets [ ], to index vectors.

Thus the MATLAB command to read the fourth value in vector v and set variable x equal to it is:

```
x = v(4)
```

To set the second value in vector v to π, type:

```
v(2) = pi
```

You can retrieve multiple values at once by providing not a single index, but a vector of indices. For example, to retrieve the 3rd through 7th values of v = 11:21, type:

```
v(3:7)
```

This will return vector [13 14 15 16 17]. Do you see why? The 3:7 created a vector of indices, and that vector was used as an index to retrieve those values of v. You can also set multiple values at once by using vectors both to index your main vector and to set those values. For example, you can type:

```
v(3:5) = [1 pi sqrt(2)]
```

This sets v(3) to 1, v(4) to π and v(5) to $\sqrt{2}$.

### *Using the End Operator*

To specify the first elements (say, the first five elements) of a vector v, you might type this notation:

```
v(1:5)
```

But how can you specify the elements from the third through the last? You could find the value of the last element using `length(v)`, and use this to build up the command:

```
v(3:length(v))
```

MATLAB provides a shortcut for this operation: the end operator, indicating the last value of the vector. To use the end operator, simply type:

```
v(3:end)
```

Similarly, the last five elements can be extracted or set using the command:

`x = v(end-4:end)` to extract, or:

`v(end-4:end)=[1 2 3 4]` to set.

### *Removing Elements in Vectors*

What does "removing an element" in a vector mean? We already know how to set it to zero; for example, setting the first element in vector v = [1 2 3] to 0 can be done with the following command:

```
v(1) = 0;
```

To fully remove the value means to shorten the length of the vector, and this is accomplished by setting the value equal to [ ], the null vector. In the above example, typing:

`v(1) = [ ]` changes vector v from [1 2 3] to [2 3].

Now `length(v)` returns `2`.

### *Building Vectors from Other Vectors*

Vectors can be assembled from other vectors and scalars. If x = [1 2 3] and y = [6 7 8], then typing:

```
z = [x 7.5 y]
```

will create vector z with the components [1 2 3 7.5 6 7 8].

### ***Inserting Elements in Vectors***

To insert a 5 at the start of vector v = [1 2 3], type the command:

```
v = [5 v]
```

You will find that vector v now holds [5 1 2 3].

To insert a 7 at the end of vector v = [1 2 3], type the command:

```
v = [v 7]
```

Vector v now holds [1 2 3 7].

To insert a 5 into the middle of vector v = [1 2 3 4], type the command:

```
v = [v(1:2) 5 v(3:4)]
```

While this command may seem like a syntactic mouthful, it follows from the preceding steps. v(1:2) returns the first two elements of v, v(3:4) returns the last two elements of v, and those with the middle inserted value of 5 are all assembled into a new vector, assigned to v. It could have been assigned to a new vector called x if the original vector v was needed in other calculations.

*PRACTICE PROBLEMS*

Given x=20:30, write the MATLAB commands to:

8. Find out how many numbers vector x holds. ©
9. Display the contents of x. ©
10. Retrieve the second value of the x vector. ©
11. Set vector y to equal the 2nd through 5th elements of x. ©
12. Set vector y = x but with a 1 inserted into the front and end of vector x. That is, y = [1 20 21 22 … 29 30 1], but create y using the x vector. ©

*PRO TIP: BUYING A DIGITAL MULTIMETER (DMM)*

The digital multimeter is an important tool for practicing electrical engineers, and one you will use frequently. Unlike relatively expensive scopes, a good DMM costs relatively little and will last a long time. If you decide to purchase your own DMM, look for the following features (in addition to the resistance and AC/DC voltage and current measuring that all DMMs should have):

- Sound continuity (buzzes when the probes are connected so you can check for shorts and opens while looking at your work—very helpful for debugging!
- Diode checking
- Capacitance measuring
- Frequency measurement (optional)

Other features are more expensive, and you are less likely to need them as you start out. These features include:

- True RMS measurement
- NIST-traceable calibration
- Go/no-go testing
- Bargraph displays

Name brand instruments like Fluke tend to be very rugged and accurate, but they also cost more. One brand used by the authors, shown here, is about 5 years old. Instruments with similar capabilities now cost less than a high-end student engineering calculator.

## COMPLEX NUMBERS

Although mathematicians use the variable i to represent $\sqrt{-1}$, electrical engineers use j since i is reserved for current (from a French word for intensity). MATLAB uses both i and j variables to represent imaginary numbers. This text will follow the EE convention of using j.

A complex number like 4 + j3 is composed of both real and imaginary parts, and it is often drawn on the complex plane, shown below:

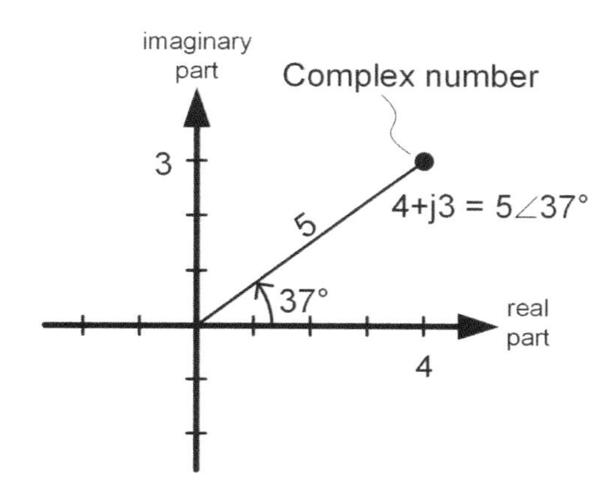


There are three common ways of writing complex numbers: rectangular, polar, and complex exponential.

### *DIGGING DEEPER*

Since a positive multiplied by a positive is positive, and a negative multiplied by a negative is also positive, it is not possible to find two real numbers that when multiplied together make a negative number. We therefore must define $\sqrt{-1}$ as something that is not real; we call it j so that $j^2 = -1$. It is ironic that one of history's great mathematicians, Euler, discovered several important relationships involving complex numbers, including the famous Euler's Identity of $e^{j\theta} = \cos(\theta) + j\sin(\theta)$, but ultimately dismissed them, saying they were mathematical curiosities of ultimately no practical significance. Now they are used throughout signal processing, communication theory, and control theory.

### *Rectangular*

This may be the most familiar to you. The complex number **z** in the diagram (above) is written as the sum of its real part and its imaginary part, or **z** = 4 + j3. Mathematicians would write the same number as 4 + 3i; electrical engineers place the j in front of its coefficient for emphasis. Notice that complex variables are bolded.

### *Polar*

The same complex number **z** can also be written as its magnitude, or distance away from the origin, and the angle that a line drawn from the origin to the complex number **z** makes with respect to the positive horizontal axis. See the diagram (opposite) for a graphical example of **z** = $5\angle 37°$, or $5\angle 0.64$ radians. Note that this represents the same number as 4 + j3 (which is expressed in rectangular coordinates).

### *Complex Exponential*

Euler's Identity paved the way for a third representation of complex numbers, closely related to polar, but which is actually a function that can be evaluated. Using the above example **z** = $5e^{j37°}$, or **z** = $5e^{j0.64}$ in radians. Note the similarity to this and the polar form.

### *Entering Complex Numbers in MATLAB*

MATLAB uses rectangular notation to enter and display complex numbers. Thus, to create a complex number like 4 + j3 in MATLAB, recall that implied multiplication is not supported, and instead type the expression:

```
z = 4 + j*3
```

### *Real and Imaginary Parts*

The real part of a complex number can be isolated by typing:

```
real(z)
```

Similarly, the imaginary part of a complex number can be isolated using:

```
imag(z)
```

### *Converting from Rectangular to Polar Form*

Polar form is written with a magnitude and angle. MATLAB cannot write the entire number **z** in polar form, but it can extract the magnitude and angle components. The magnitude of a complex number **z** can be found by typing:

```
abs(z)
```

The angle in radians is:

```
angle(z)
```

Convert to find the angle in degrees by multiplying by 180/π:

```
angle(z)*180/pi
```

Once the magnitude and angle components are found, the number can be handwritten in polar notation using the format $\text{magnitude} \angle \text{angle}$ or $\text{magnitude} \angle \text{angle}^\circ$ if in degrees, e.g. $5\angle 37^\circ$.

### *Converting from Polar to Rectangular Form*

You can use trigonometry and the complex plane diagram (shown earlier in this chapter) to derive the following equations that convert polar to rectangular form, allowing you to enter polar-notated numbers into MATLAB.

Given the following complex number in polar form, **z** = mag ∠ ang, enter it in MATLAB as follows:

```
z = mag*cos(ang) + j*mag*sin(ang)
```

if the angle `ang` is in radians, or:

```
z = mag*cosd(ang) + j*mag*sind(ang)
```

if the angle `ang` is in degrees.

### ***Converting to and from Complex Exponential Form***

It is possible to convert a complex number in exponential form like **z** = mag $e^{j\theta}$, with θ in radians, into rectangular form by simply entering it directly into Matlab, since Matlab displays complex numbers in rectangular form. For example, $3e^{j\pi}$ can be directly entered as:

```
z = 3*exp(j*pi);
```

If the angle is given in degrees, it can be converted into radians by multiplying by π/180. For example, $7e^{j22^\circ}$ can be entered as:

```
z = 7*exp(j*22*pi/180);
```

To convert a number in complex rectangular form to complex exponential form, first convert it to polar form as described earlier, and then write the magnitude and angle components in this form:

**z** = mag $e^{j\theta}$

For example, let **z** = 4 + j3. Converted to polar notation, this is is 5∠0.64, which in complex exponential notation is then written as $5e^{j0.64}$.

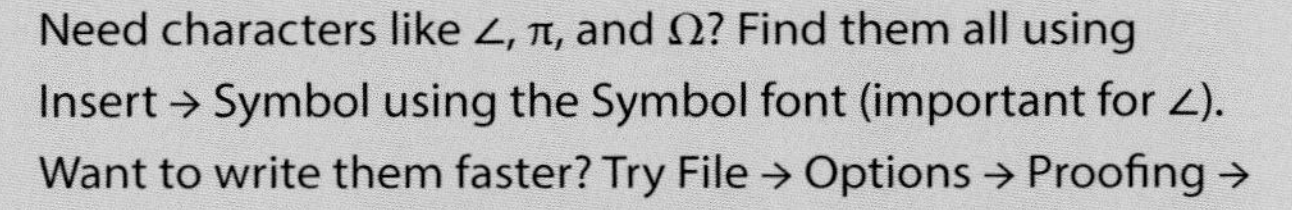

NOTE

Need characters like ∠, π, and Ω? Find them all using Insert → Symbol using the Symbol font (important for ∠). Want to write them faster? Try File → Options → Proofing → Autocorrect → Autocorrect Options → Math Autocorrect → Check the "Use Math Autocorrect Outside of Math Regions" option. Then in your document, type \angle to get ∠, \pi to get π, and \Omega to get Ω. Note that these commands will only work in the desktop version of MS Word, not the online version.

### PRACTICE PROBLEMS

13. Write the complex number `z = 5+j12` in polar form with the angle in degrees. ©®
14. Write the same complex number `z = 5+j12` in complex exponential form with the angle in radians. ©®
15. Use MATLAB to convert $14\angle 45°$ into rectangular form. ©®

## VECTOR MATHEMATICS

Many functions take both scalar and vector arguments, and with one important difference, these work as you would expect: functions that take a single argument, like `cos()`, operate on each element of the vector independently. Operators that take two vectors, like +, operate on respective elements of each vector. For instance, if `x = [2 4 7]`, then `cos(x)` returns a vector of all three cosines and `x + x` returns `[4 8 14]`. Other functions that you have used that take vector arguments include: addition, subtraction, `sqrt()`, `exp()`, and all the trigonometric functions like `sin()`.

Vector mathematics using the four basic operations of addition, subtraction, multiplication, and division have some special behaviors. Addition and subtraction only operate on respective elements: `[5 3]-[1 2] =[4 1]`. **Multiplication and division between two vectors do not use the standard * and / operators.** For reasons described in linear algebra courses, multiplying two row vectors like `[5 3]*[1 2]` will generate an error message. To multiply or divide respective elements, prefix the multiplication or division sign with a dot. For example, `[5 3].*[1 2]` will work, as will `[4 8]./[2 4]`, yielding `[5 6]` and `[2 2]`, respectively.

All four basic operators work when one argument is a scalar and one is a vector. For example:

```
[2 4 9] - 1
```

returns

`[1 3 8]`, just as

`[2 4 9] - [1 1 1]` does.

`[2 8 16]/2` is the same as

`[2 8 16]./[2 2 2]`

both of which yield

```
[1 4 8]
```

*PRACTICE PROBLEMS*

16. Given `x = [4 2 6]`, write the MATLAB command to add 5 to each element. ©
17. Given `x = [4 2 6]` and `y = [2 4 8]`, write the MATLAB command to multiply corresponding elements together. If MATLAB returns an error, re-read the bolded section above. ©

## STRINGS

Although strings are not strictly a calculator-like function, MATLAB treats strings as vectors of characters.

To enter a string, enclose it in **single** quotation marks, as follows:

```
s1 = 'Hello, there!';
```

To concatenate strings, or set/retrieve individual letters or substrings, use the same commands as if working with numeric vectors. For example:

- To retrieve the second letter of `s1` defined above, an "e", type:

  ```
  s1(2)
  ```

- To return the first word "Hello" in the string, type:

  ```
  s1(1:5)
  ```

- To insert the word "Laura" before the exclamation point, use the command:

  ```
  [ s1(1:12) ' Laura' s1(13)]
  ```

- To delete everything except the word "Hello," type:

  ```
  s1(6:14) = [];
  ```

### *Converting Between String Representations of Numbers and Actual Numbers*

One could define a string representing a number like:

```
string2 = '521';
```

While `string2` looks like a number, it is a string, which is really a vector of characters. This is apparent by adding 0 to it, like this:

```
string2 + 0
```

Instead of returning 521, it returns the vector `[53 50 49]` which are the ASCII numbers representing the characters `['5' '2' '1']`.

To convert a string representation of a number into the number, use `str2num`:

```
str2num('521') + 3
```

now returns the number `524` as expected.

PRACTICE PROBLEMS

18. If `str = 'R1 = 47 ohms'`, write the MATLAB command that isolates the number 47 from the string, converts it into a number, and stores it in variable R1. ©

## *TECH TIP: RESISTORS IN SERIES AND PARALLEL*

The most common electronic circuit component is the resistor. Our analogy in the Ch. 1 Tech Tip on page 11 likened resistors to a restriction in a pipe that limits the flow of water (current) when driven by a constant pressure (voltage). Two resistors of resistance $R_1$ and $R_2$ can be connected in series or parallel to create a new equivalent resistance, as shown below:

| | **Two Resistors** | **Equivalent Resistance** | **Water Analogy** |
|---|---|---|---|
| **Series** | $R_1$ $R_2$ | $R_1+R_2$ | $R_1$ $R_2$ |
| **Parallel** | $R_1$ $R_2$ | $\frac{R_1 \cdot R_2}{R_1+R_2}$ | $R_1$ $R_2$ |

It is important for EE students to develop their intuition so that the mathematical equations become obvious. Using the water analogy, think of two resistors in series as two restrictions in series, as shown above. A pump (voltage source) must push water (current) through the first, and then through the second, so it follows that the equivalent resistance is the sum of the two individual resistances, or $R_1 + R_2$. The water analogy for two resistors in parallel shows that adding a second resistor increases the number of paths that water can flow, so it decreases resistance, much as opening a second adjacent door will decrease resistance to traffic flow. The equivalent resistance of two resistors in parallel is:

$$\frac{R_1 \times R_2}{R_1 + R_2}$$

This formula is often memorized as "product over sum," and will always return an equivalent resistance lower than either of the resistors that comprise it.

### *PRACTICE PROBLEMS*

19. Write the MATLAB command that, given variables $R_6 = 60$ and $R_7 = 120$, computes their values in parallel. Use the variables $R_6$ and $R_7$ in your MATLAB command, not the numbers they hold (for example, in series, the MATLAB command is `R6+R7`). ©

## MATRICES

Thus far, the vectors discussed in this chapter have been either a single column or row of numbers. A **matrix** is a collection of both rows and columns of numbers, e.g.:

$$\begin{bmatrix} 2 & -4 & 1.7 \\ 42 & \sqrt{2} & 0 \end{bmatrix}$$

The dimensions of the matrix indicate how many rows and columns it has, with the number of rows listed first, followed by the number of columns. Thus the 2-row x 3-column matrix above is a 2x3 matrix. Matrices are used in digital signal processing to hold streams of data, where the data streams can be very large. For example, a 100-second song recorded at the CD-quality of 44,000 samples/second can be represented as a 2-column x 4,400,000-row matrix. The rows represent the left and right channels of the signal, and each column represents a sample, i.e. a snapshot in time of the voltage to be applied to a speaker.

### *Special Matrix Dimensions*

If a matrix has a single row, such as 1x5, it is called a **row vector**. The vectors discussed earlier in this chapter are examples of row vectors. A **matrix** may also have a single column, like 5x1. These are called **column vectors**. Matrices may also have a single row and column; these matrices are **scalars**, or "regular" numbers. Matrices may even be **null**, meaning they have zero rows and columns; then they hold nothing.

## CREATING MATRICES

As with creating vectors, there are several different ways to create matrices:

- Defining components explicitly
- Filling with duplicate values
- Filling with vectors
- Filling with random values

### *Defining Components Explicitly*

To create the matrix x = $\begin{bmatrix} 2 & -3 \\ 1 & 0 \end{bmatrix}$

type the following in MATLAB:

```
x = [2 -3; 1 0]
```

Compare that with how you earlier defined the top row vector y = [2 -3]; expressed as follows:

```
y = [2 -3]
```

The semicolon operator ; separates rows when used inside a matrix definition. Note that square brackets `[]` are used to define matrices and vectors.

*RECALL*

The semicolon has another purpose besides separating matrix rows; when placed at the end of a line, it prevents MATLAB from displaying the calculation results.

### *Filling with Duplicate Values*

The `zeros()` and `ones()` commands can work to create matrices, as well as vectors as described previously.

`zeros(2,4)` creates a matrix filled with 2 rows of 4 zeros, and is equivalent to typing:

```
[0 0 0 0; 0 0 0 0]
```

`ones(3,5)` creates a matrix filled with 3 rows of 5 ones, and is equivalent to typing:

```
[1 1 1 1 1; 1 1 1 1 1; 1 1 1 1 1]
```

To create a 20 x 30 matrix called `big` filled with 7's, use the fact that MATLAB understands multiplying a matrix by a scalar, like so:

```
big = 7 * ones(20,30);
```

The semicolon at the end of this command is useful; without the semicolon, the screen would fill with the 600 copies of the number 7 that it generates.

*NOTE*

Notice the link between these matrix definitions and vectors you earlier defined. When you created a vector of five repeating 1's by using `ones(1,5)`, you were actually creating a matrix of 1 row and 5 columns.

***Filling with Vectors***

Vectors can be combined to form matrices. Earlier we described how to concatenate vectors `x = [1 2 3]` and `y = [4 5 6]` into one long row vector `z` by typing `z = [x y]`. Using the semicolon instead of a space, i.e. `z = [x; y]`, will create the matrix

$z = \begin{bmatrix} 1 & 2 & 3 \\ 4 & 5 & 6 \end{bmatrix}$, just as typing `z = [1 2 3; 4 5 6]` would.

***Filling with Random Values***

Matrices filled with random numbers between 0 and 1 may be created using the `rand()` function. For instance, to generate a matrix called `mrand` of 3 rows and 3 columns, filled with 9 numbers varying between -1 and +1, type the following:

```
mrand = 2*rand(3,3) - 1;
```

### *PRACTICE PROBLEMS*

Write the MATLAB commands to generate the following matrices:

20. Matrix with the first row containing -2 and 3, and the second row containing 100 and 0. ©
21. Matrix of 3 rows and 5 columns filled with 0's. ©
22. Column matrix of 10 rows and 1 column filled with 2's. ©
23. Assume you have row vectors R1 and R2. You do not know what the elements of R1 and R2 are, but you know they are each different row vectors of length 8. Write the commands to create a 2-row by 8-column matrix made using R1 and R2. ©
24. Create a 100 row by 2 column matrix using the pre-defined column vectors C1 and C2, where C1 and C2 are predefined to be column vectors of length 100. ©

## CHANGING MATRIX VALUES

### ***Accessing and Changing Values in Matrices***

Matrices are addressed using rounded parentheses; always in row, column order; and always using the convention that the row and column index counting begins with 1. For example, if:

`m = [-5 7 2.4; 8 0 9]` ($x = \begin{bmatrix} -5 & 7 & 2.4 \\ 8 & 0 & 9 \end{bmatrix}$)

then:

```
x = m(2,3)
```

will set the scalar variable x equal to 9.

`m(2,3) = pi` creates the following matrix:

$$\begin{bmatrix} -5 & 7 & 2.4 \\ 8 & 0 & 3.14159 \end{bmatrix}$$

### *Accessing and Changing Whole Rows and Columns*

Much like if:

`v = [-2 -4 -6 -8 -10]` then

`v(2:4)` returns

`[-4 -6 -8]`, we can address parts of matrices using the colon `:` operator.

Using the example matrix m from the previous section, where m = $\begin{bmatrix} -5 & 7 & 2.4 \\ 8 & 0 & 9 \end{bmatrix}$

`m(1:2, 1:2)`

returns the square matrix $\begin{bmatrix} -5 & 7 \\ 8 & 0 \end{bmatrix}$

The colon operator is even more powerful than this. It can be used to specify an entire row or column using the shortened form: `m(:, :)`. For example, to set x equal to the second row of m, type:

`x = m(2, :);`

Now x = [8  0  9]

To set the first column of m to $\begin{bmatrix} 1 \\ 2 \end{bmatrix}$ (note how a single column of a matrix is a column vector), type:

`m(1,:) = [1;2];`

Now m = $\begin{bmatrix} 1 & 7 & 2.4 \\ 2 & 0 & 9 \end{bmatrix}$

### *Removing Elements in Matrices*

Similar to vector manipulation, remove chunks of matrices by setting them equal to the null matrix `[]`. Note that you can delete any number of whole rows or columns of a matrix, but unlike with vectors, it does not make sense to delete a single value; then the matrix would not be square or rectangular. For example:

| | |
|---|---|
| `m(:,4) = []` | deletes the 4th column |
| `m(2:3,:)=[]` | deletes rows 2 and 3 |
| `m(:, 3:end)=[]` | deletes all columns except the first two |
| `m([1:2:end,:])=[]` | deletes all odd-numbered rows |

## PRACTICE PROBLEMS

Use the matrix `m` for these problems. You do not know what the elements of `m` are, but you know it is a 2x2 matrix.
Write the MATLAB command to:

25. Create variable `x` and set it equal to the second row, second column value of `m`. ©
26. Set the 2nd row, 1st column value of `m` to 0. ©
27. Remove the entire top row of `m`. ©
28. Replace the first column of `m` with random numbers created using `rand()`. ©
29. Create a vector `v` equal to the last column of `m`. ©

### PRO TIP: BUYING A CALCULATOR

Electrical engineers in industry have their own offices, and typically perform computations using MATLAB. But for college students, one of the most important tools is the portable calculator. Unlike most majors, ECE students must have calculators capable of handling complex matrices to solve problems involving circuits driven by sinusoidal sources; the TI-86 and below will not do this, including the TI-84. Don't be mislead into purchasing an expensive instrument with many specialized engineering programs built into it; these are rarely useful. A TI-89, TI-91, or higher, or the new TI-Nspire series all work well. The Nspire's document-centric nature is unique; some students prefer it but many see it as an abstraction that separates them from simply solving equations.

## WORKING WITH MATRICES

### *Matrix Size*

Just as the `length()` command can be used to determine how many elements are in a vector, the `size()` command returns how many rows and columns are in a matrix. For example, if:

`x = [4 3; 2 -4; 5 7]` (that is, x= $\begin{bmatrix} 4 & 3 \\ 2 & -4 \\ 5 & 7 \end{bmatrix}$),

then the matrix dimensions can be using:

```
[rows, cols] = size(x)
```

This sets the variable `rows` to 3 and `cols` to 2. Note how `size()` sets two variables at the same time. Since a vector is just a matrix with one row or column, `size()` also works on vectors:

`size([10 20 30 40 50])` returns [1, 5] and

`size([10; 20; 30; 40; 50])` returns [5, 1].

### *Transpose*

Transposing a real matrix interchanges rows for columns. Thus the matrix on the left in the example below

$\begin{bmatrix} 4 & 3 \\ 2 & -4 \\ 5 & 7 \end{bmatrix}$ becomes $\begin{bmatrix} 4 & 2 & 5 \\ 3 & -4 & 7 \end{bmatrix}$ when transposed.

The MATLAB command to do this is the apostrophe '. For example:

`x = [4 3; 2 -4; 5 7]` enters the matrix in the upper left, and

`y = x'` creates the matrix in the upper right.

This can also be used to change column vectors to row vectors and vice-versa. For example, defining the column vector:

```
resistors = [10; 11; 12; 13; 15]
```

is the same as defining the transpose of the row vector

```
resistors = [10 11 12 13 15]'
```

and both define the column vector $\begin{bmatrix} 10 \\ 11 \\ 12 \\ 13 \\ 15 \end{bmatrix}$

***Matrix Mathematics***

Matrix mathematics using the four basic operations of addition, subtraction, multiplication, and division have special behaviors according to the rules of matrix algebra. Addition and subtraction operate on respective elements:

$$\begin{bmatrix} 1 & 5 \\ 2 & 3 \end{bmatrix} + \begin{bmatrix} -1 & 6 \\ 0 & 1 \end{bmatrix} = \begin{bmatrix} 0 & 11 \\ 2 & 4 \end{bmatrix}$$

However, as you learned with vectors, **multiplication and division between two matricies do not operate on respective elements using the * and / operators.** If this is needed, prefix the multiplication or division sign with a dot.

For example, `[2 15; 0 3].*[-1 3; 2 1]` will work as expected, yielding `[-2 45; 0 3]`. Multiplication and division do work with the standard `*` and `/` operators between a matrix and a scalar, where it is understood each element of the matrix will be multiplied or divided by that scalar. For example, `[2 -3; 0 4]*2` yields `[4 -6; 0 8]`.

## PRACTICE PROBLEMS

30. What single MATLAB command will allow you to evaluate the expression below given the vector of t values `t=0:5`? ©®

$$y = \frac{\cos\left(\frac{t}{2} + \frac{\pi}{4}\right)}{t+1}$$

*TECH TIP: MESH AND NODAL ANALYSIS METHODS*

Techniques for solving complex circuits with many resistors include Node Voltage Analysis and Mesh Current Analysis. These methods are usually taught in circuit analysis courses, but at their core, they are a systematic way of applying Ohm's Law, ***V = I R,*** repeatedly over circuit fragments. They create systems of equations that can be solved simultaneously to find every voltage (for Node Voltage Analysis) or current (for Mesh Current Analysis) in the circuit. MATLAB can solve these simultaneous equations, which sometimes include more than 100 unknowns, in less than a second.

The diagrams below show on the left, a circuit, its voltage definitions, and the set of equations generated by the Node Voltage Analysis method for its voltages; and on the right, the same circuit with its currents defined, as well as the set of equations derived using the Mesh Current Analysis method to find these currents. Beginning EE students should know the names of these methods, but need not know how to apply them yet.

**Node Voltage Analysis**

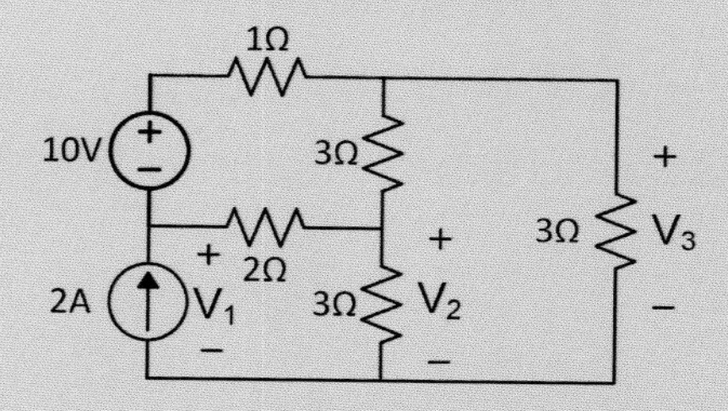


voltage equations for above

$$9V_1 - 3V_2 - 6V_3 = 12$$
$$-3V_1 + 7V_2 - 2V_3 = 60$$
$$-3V_2 + 10V_3 = 60$$

**Mesh Current Analysis**

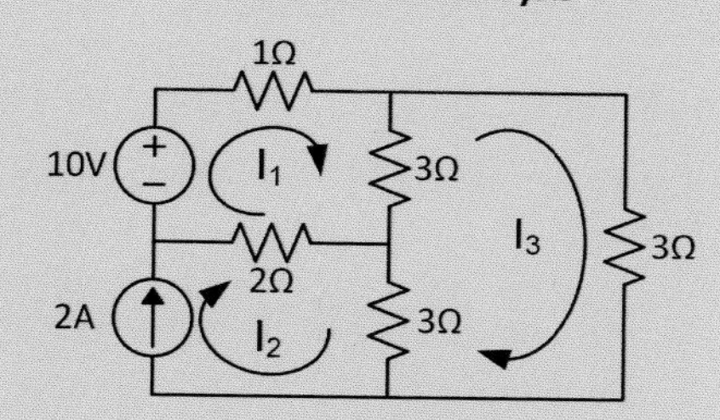


current equations for above

$$6I_1 - 2I_2 - 3I_3 = 10$$
$$I_2 = 2$$
$$-3I_1 - 3I_2 + 9I_3 = 0$$

## SOLVING SIMULTANEOUS EQUATIONS USING MATRICES

Since nodal analysis and mesh analysis are very common methods of circuit analysis and yield sets of linear real or complex simultaneous equations, solving simultaneous equations is a basic staple of the electrical engineering curriculum. In MATLAB this is accomplished as shown in the following example.

Solve for the various voltages $V_1$, $V_2$, and $V_3$ in the following set of equations:

$$\begin{aligned} V_1 + 2V_2 + V_3 &= 5 \\ 3V_1 - V_2 + 2V_3 &= 2 \\ -V_1 + V_2 - 2V_3 &= -4 \end{aligned}$$

To do this, create an ***A*** matrix with the coefficients of the unknown voltages and a ***b*** column vector with the values to the right of the equals sign, as follows:

```
A = [1 2 1;3 -1 2;-1 1 -2];
b = [5; 2; -4];
```

The solution is (note the backwards division sign):

```
V = A\b
```

When using nodal and mesh methods to analyze the response of circuits containing ***capacitors*** and ***inductors*** you will solve sets of complex matrices. There is no difference as far as MATLAB is concerned; the ***A*** matrix will have complex values such as for this example solving for complex currents:

```
A = [1+j  2-j*2; -3+j 3];
b = [j; 2-j];
I = A\b;
```

On a vintage 2012 computer, MATLAB can solve a set of 100 equations with 100 unknowns in about 1 ms—less time than it takes to depress the Enter key fully.

### DIGGING DEEPER

Matrices can do more than solve simultaneous equations for circuit analysis; they are essential to manipulating video, rotating photos, providing equalization to audio streams, finding particular frequencies present in signals, controlling servo motors, and are the discrete mathematics counterpart of integration and differentiation.

## PRACTICE PROBLEMS

31. Use MATLAB to solve the following set of mesh equations for a circuit to find $I_1$, $I_2$, and $I_3$: ©®

$$2I_1 + 6I_2 - 3I_3 = 3$$
$$4I_1 - I_2 + I_3 = 25$$
$$I_1 + 2I_2 - I_3 = 8$$

### TECH TIP: VOLTAGE DIVIDERS

One of the most common circuits is the voltage divider. It is composed of two resistors in series so that a voltage applied across both is split, with part being dropped by the first resistor and the remainder by the second as shown in the schematic below. The formula to find the voltage $V_1$ that is across $R_1$ is given in the figure:

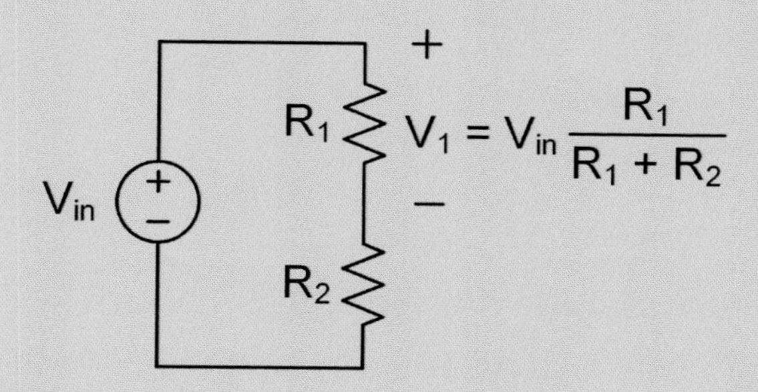


As with all new ECE formulae, take a moment to analyze it:

- The voltage drop across $R_1$ is proportional to the applied voltage $V_{in}$ which makes sense; if one doubles the applied voltage, the voltage across each of the resistors should double.

- The amount of voltage dropped by $R_1$ is proportional to the ratio of $R_1$ to the total circuit resistance $R_1 + R_2$. Since the entire applied voltage must be dropped by the two resistors, it is intuitive that it is split according to each resistor's relative contribution to the total resistance.
- By the same reasoning, the voltage across $R_2$ is

$$V_2 = V_{in}\frac{R_2}{R_1 + R_2}$$

This means the sum of the voltage drops across both resistors is:

$$V_{in}\frac{R_1}{R_1 + R_2} + V_{in}\frac{R_2}{R_1 + R_2}$$

which simplifies to $V_{in}$, the total applied voltage, just as we expect.

*PRACTICE PROBLEMS*

32. Use the circuit given in the above Tech Tip with $R_1 = 15\Omega$ and $R_2 = 6\Omega$, use MATLAB and algebra to find the $V_{in}$ needed to make $V_1 = 19V$. ©®

## COMMAND REVIEW

### *Creating Matrices*

`[]` defines each element of a matrix, e.g. `m = [2 4; 1 5]`

`zeros(r,c)` creates a matrix of zeros with r rows and c columns

`ones(r,c)` creates a matrix of ones with r rows and c columns

`[a:b]` creates a vector from a to b incrementing by 1

`[a:inc:b]` creates a vector from a to b incrementing by inc

`linspace(a,b,N)` creates a vector from a to b with N equally-spaced elements

`logspace(a,b,N)` creates a vector from $10^a$ to $10^b$ with N exponentially-spaced elements

`rand(r,c)` creates a matrix with r rows and c columns of random numbers from 0 to 1

### *Rounding Numbers*

`round(n)` rounds to the nearest integer, so [-3.6 5.7] becomes [-4 6]

`floor(n)` rounds down to the next integer, so [-3.6 5.7] becomes [-4 5]

`ceil(n)` rounds up to the next integer, so [-3.6 5.7] becomes [-3 6]

### *Querying Matrices*

`length(v)` returns the number of elements in vector v

`[r,c]=size(m)` returns the number of rows in r and columns in c

`m(r,c)` returns the element of m at row r and column c. r and c may be vectors.

`m(:, c)` returns the column c of m (i.e. all rows, column c)

`m(1:4, :)` returns rows 1 through 4 of m (i.e. rows 1 to 4, all columns)

`m(:, 3:end)` returns columns from 3 onwards of m (i.e. all rows, columns 3 to end)

### *Deleting Parts of Matrices*

`v(4) = []` removes element 4 of vector v

`m(:,1) = []` removes the 1st column of m

### *Mathematics*

`pi` π

`+ - * /` work between two scalars or a scalar and matrix. No implied multiplication.

`+ -` work between the corresponding elements in two matrices to add and subtract.

`.* ./` work between the corresponding elements in two matrices to multiply and divide.

`A\b` solves the set of *n* simultaneous equations where A is an *n* x *n* matrix of coefficients and b is a *n* x 1 column vector

### *Complex Numbers*

`i, j` sqrt(-1), can be used for instance as 3 + j*7;

`real(z)` the real part of complex number **z** (rectangular coordinates)

`imag(z)` the imaginary part of complex number **z** (rectangular coordinates)

`abs(z)` the magnitude of complex number **z** (polar coordinates)

`angle(z)` the angle of complex number **z** in radians (polar coordinates)

### *Strings*

`v = 'Hello'` defines a string

`x = str2num('42')` changes a string representation of a number to a number

## LAB PROBLEMS

© = Write only the MATLAB command
® = Write only the MATLAB result

1. Resistors are available in certain pre-defined sizes, so you can't expect to find one with a value of exactly, say, π Ohms. Write the MATLAB commands that create the vector v1 of all **standard** 5% resistor values from 1 Ω through 9.1 Ω. The list of standard values is described in the Tech Tip titled "Measuring Resistance" on page 38 ©®
2. Create vector v2 by modifying v1 (above) so it prints all standard 5% resistor values from 1 Ω through and including 1 MΩ (same as 1,000,000 Ω). You will use these vectors again, so save both v1 and v2 in the MATLAB file, resistors.mat. Hint: How would you scale v1 to make it 10 times bigger? How would you then concatenate v1 and that scaled v1? Make sure the last value is 1 MΩ, not 910,000Ω. ©
3. If you are given a vector x, write the MATLAB commands to:
    a) extract the first 5 numbers ©
    b) delete the last 5 numbers ©
4. Make a vector v3 of the alternating digits [1 -1  1 -1...] that is $2^{10} = 1024$ long. Hint: Do not specify each value independently, since that would take too long. Instead, either think about how to concatenate [1 -1] into longer and longer units, or alternatively, think about how to use the fact that $(-1)^0 = 1$, $(-1)^1 = -1$, $(-1)^2 = 1$, $(-1)^3 = -1$, etc. ©
5. Find the equivalent parallel resistance of a 5 Ω resistor placed in parallel separately with each of the standard-value resistors from 1 Ω through 9.1 Ω. (You already created that vector in Problem 1).
   That is, find the equivalent of a 5 Ω resistor in parallel with a 1 Ω resistor, and repeat that with the 5 Ω in parallel with a 1.1 Ω resistor, etc. up to the 5 Ω in parallel with a 9.1 Ω resistor. Do this in MATLAB by defining R1 = 5 and R2 = the vector of resistances you defined in Problem 1. Then define the single equation, and MATLAB's result, that answers the question. ©®

6. Let $\mathbf{z} = \dfrac{2 + j7}{23\angle 45^{\circ}}$

   Use MATLAB to calculate **z** and display in both rectangular and polar notation. Use degrees for the angle in polar notation. ®

7. Create a string variable called str1 = '341'. Type the MATLAB command that isolates the last two characters (that is, 41), converts them into a number, and adds 1 to the result. ©

8. A mesh analysis of a circuit yields the equations listed below. Use MATLAB to solve them for $I_1$, $I_2$, $I_3$, and $I_4$. Hint: If an equation does not mention a variable, there is 0 times that variable. ©®

   $$2i_1 - i_2 + i_3 + 6i_4 = 20$$
   $$-i_1 + 3i_2 = -12$$
   $$i_3 + i_4 = 1$$
   $$2i_1 - i_3 = 12$$

9. A nodal analysis of a circuit yields the following set of complex equations. Use MATLAB to find V1 and V2, and write the result in rectangular notation. ®

```
(1+j2) V1 + j3 V2 = -1+j12
     2 V1 -  4 V2 = -16+j34
```

# 3 MATLAB GRAPHICS

## OBJECTIVES

After completing this chapter, you will be able to use MATLAB to do the following:

- Create line plots, similar to an oscilloscope display
- Create multiple line plots in the same axes, like a multiple trace oscilloscope
- Create scatter plots to show individual data pairs, like calculated vs. measured voltage
- Create bar plots to show grouped data
- Create linear or logarithmically-spaced axes
- Create plots with multiple vertical axes
- Add text labels to plots, including using Greek symbols

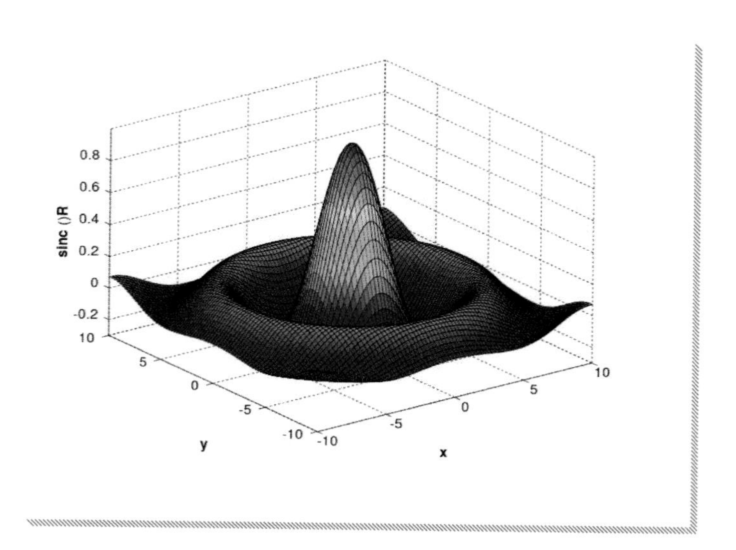

## TYPES OF PLOTS

Electrical engineers use a wide variety of graphs to show data. The most common types are the line plot, the scatter plot, and the bar plot, as shown below. The line plot is used to show continuously-varying data, much as an oscilloscope may be used to view a time-varying voltage. The scatter plot is used to show discretely-measured data (i.e. the actual vs. rated resistance of a sample of five resistors). If there is too much data to display using a scatter plot (i.e. a data set sampling 10,000 resistors), the data may be grouped into bins and displayed in a bar plot.

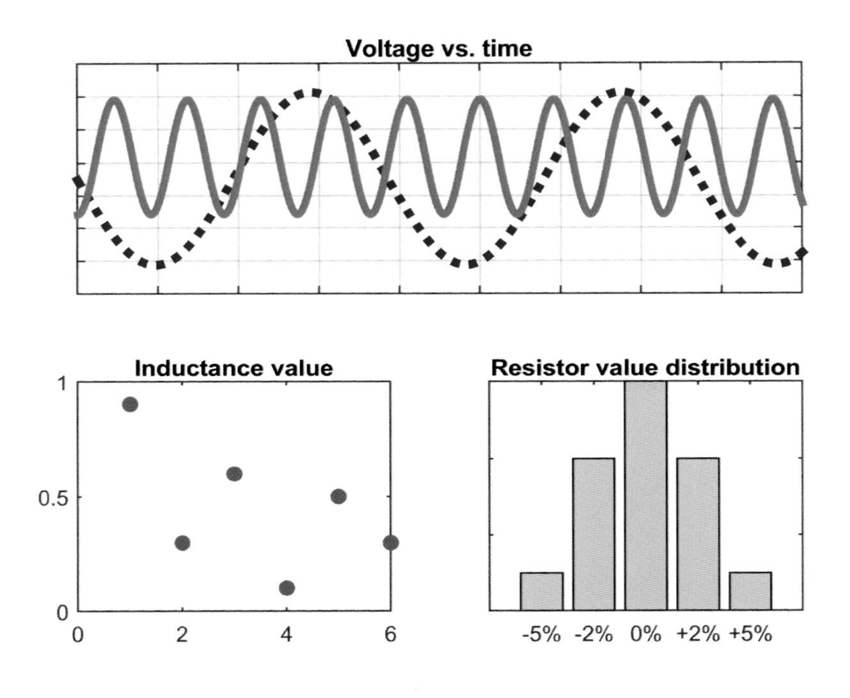


Common types of plots used in electrical and computer engineering are shown above. "Voltage vs. time" shows the line plot (with two differently-formatted lines on the same axes), while "Inductance value" and "Resistor value distribution" are examples of the scatter plot and the bar chart, respectively. In this chapter you will learn how make all of these in MATLAB.

## LINE PLOT

The line plot is the most common type of graph. It requires two vectors of equal length: one that specifies the horizontal coordinates of each point, and one that specifies the corresponding vertical points. In high-school mathematics classes, horizontal distances are often called *x* and vertical distances *y*. In electrical engineering, the horizontal axis is often *t* for time, and the vertical axis is a voltage *v*(*t*), a current *i*(*t*), or a generic signal, such as *x*(*t*) or *y*(*t*).

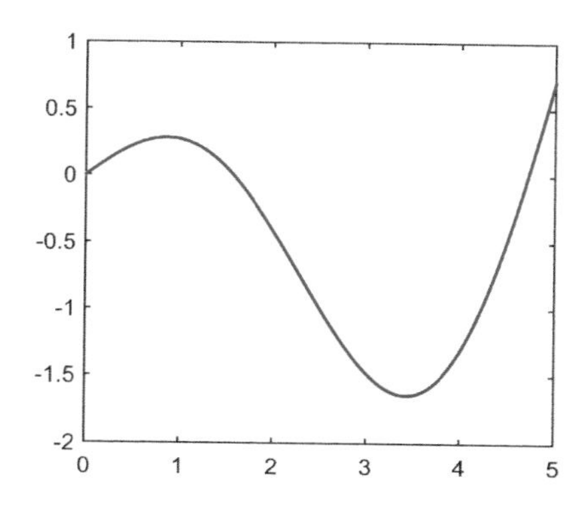


To plot the equation $y(t)=\frac{t}{2}\cos(t)$ between $0 \le t \le 5$ shown above:

1. Create the horizontal t vector with enough points to give a smooth plot:
   ```
   t = linspace(0, 5, 1000);
   ```
2. Recall that a semicolon is needed at the end of the command to suppress echoing the 1000 elements of t to the screen.
3. Create the vertical y vector corresponding to the desired function:
   ```
   y = t/2 .* cos(t);
   ```
   Remember that `.*` is needed for multiplication between corresponding vector elements.
4. Create the plot using `plot(horizontal, vertical)`
   ```
   plot(t,y)
   ```

*RECALL*

The the commands +, –, *, / work between a scalar and a vector. Between two vectors, the commands + and – add or subtract corresponding elements of the two vectors. You must use .* and ./ to multiply or divide corresponding elements between two vectors.

***Selecting Line Colors***

MATLAB will select blue for plot lines by default. To select a different color, such as plotting the previous example in red, type:

```
plot(t, y, 'r')
```

where 'r' stands for red.

MATLAB has pre-defined eight colors with single-letter shortcuts:

| Color | Abbreviation |
|---|---|
| blue | 'b' |
| green | 'g' |
| red | 'r' |
| cyan | 'c' |
| magenta | 'm' |
| yellow | 'y' |
| black | 'k' |
| white | 'w' |

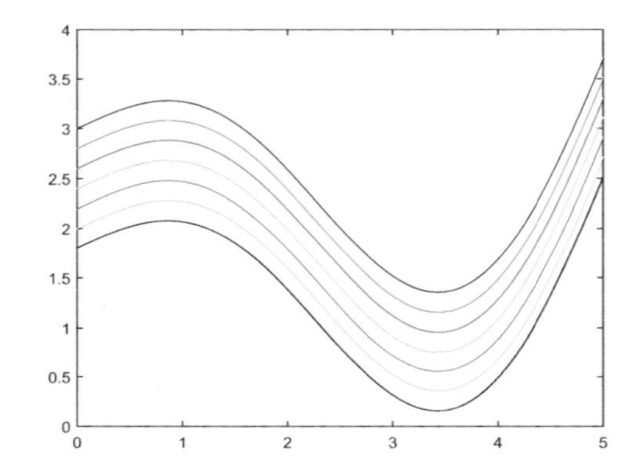


You do not need to memorize these abbreviations; find them by typing:

```
help plot
```

*DIGGING DEEPER*

If you need to embed MATLAB plots in Microsoft Word, do not use screen grab tools, which only capture low-quality bitmaps that will appear blurry when printed. Instead, select the plot window you wish to copy; choose Edit -> Copy Figure from the menu; and then paste it into the Word document.

### *Selecting Line Styles*

By default, MATLAB uses solid lines to draw plots; however, other line styles can be specified, such as dotted, dashed, and dash dot:

| Style | Abbreviation |
|---|---|
| solid | '-' |
| dotted | ':' |
| dashed | '--' |
| dash dot | '-.' |

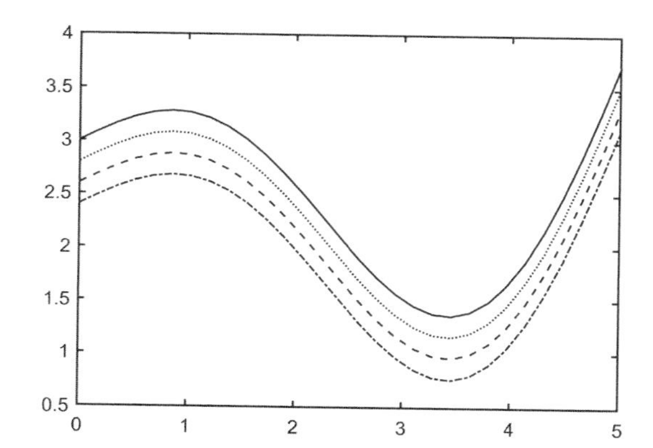


To create a dotted line, as shown in the plot to the right, type:

```
plot(t, y, ':')
```

This command can be used in addition to the line color command, as shown in the following example:

```
plot(t, y, 'r:')
```

This creates a red dotted line. Notice how the commands are combined into a single quoted command.

### *Choosing Line Width and Arbitrary Colors*

Specifying pre-defined colors and line styles is so common that they are integrated in the plot command itself. Other less-common options, such as choosing the width of the line or choosing an arbitrary color, are accessed through the keywords at the end of the plot command, followed by a number or vector.

### *Varying Plot Line Width*

The line width is set to 1 by default. To make it five times thicker, choose a line-width of 5 using this command:

```
plot(t, y, 'linewidth', 5)
```

Examples of plots with line widths ranging from 1 to 15 are shown below:

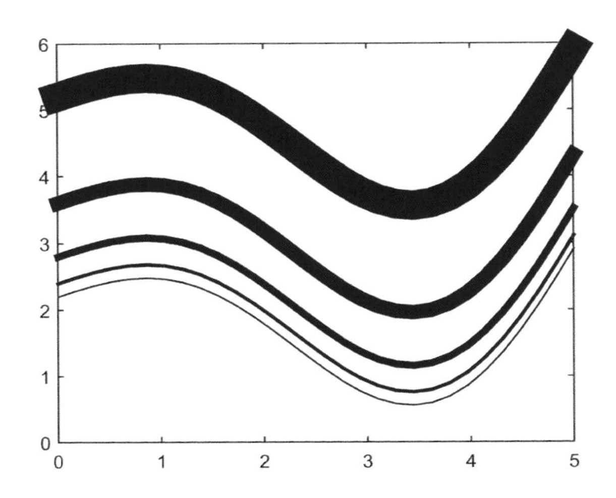


Similarly, if you need a color other than one of the eight pre-defined colors that MATLAB provides, use the `'color'` keyword, followed by the color specification in [red, green, blue] coordinates. This is a vector of three numbers, with each number varying between 0 and 1, that defines how much red, blue, and green (respectively) are present. This color specification system is a common one in computer graphics. For example, orange is made with a lot of red, mixed with a medium amount of green and no blue, so the RGB coordinates [1 0.5 0] would yield an orange hue, which can be plotted using the following command:

```
plot(t,y, 'color', [1 0.5 0])
```

### *Combining Commands*

Commands to `plot()` must be grouped into two parts. The first is the x,y data, followed by the single-character color and the one-or-two-character linestyle. Any color and linestyle characters must all be grouped together and enclosed in a single set of quotes, **e.g.** `'r-'` **rather than** `'r','-'`. Commands that are given in two parts, like `'linewidth', 10` must come at the end. For example, to draw the thick red dotted line shown on the next page, use the command:

first part　　　second part

```
plot(t,y,'r:','linewidth',10)
```

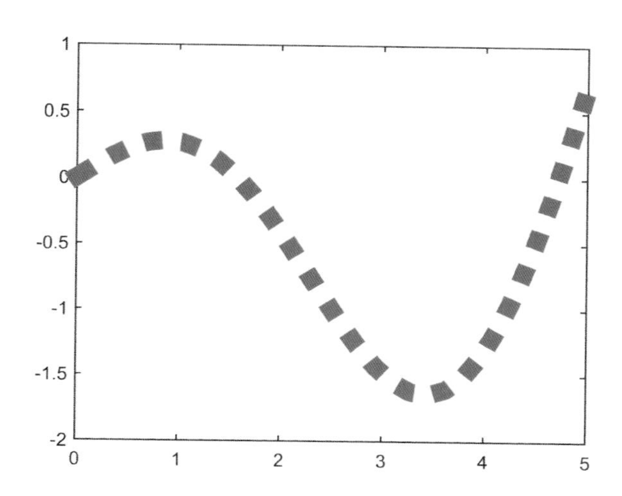


PRACTICE PROBLEMS

1. Plot cos(t) from $-2\pi \leq t \leq 2\pi$. ©®
2. Repeat the above using a dotted, green line, with a width ten times thicker than the default. ©®

### *Axis Labels and Titles*

Most plots will need a title and labels for each axis. This can be done after the plot is created using the `title`, `xlabel`, and `ylabel` commands, as shown in the example below, which plots the first five seconds of a decaying voltage exponential:

```
t = linspace(0,5,100);
y = exp(-t);
plot(t,y)
title('Decaying Exponential')
xlabel('time (s)')
ylabel('amplitude (V)')
```

Notice that strings in MATLAB are contained in single quotes, unlike the double quotes used by most programming languages. See the resulting plot below:

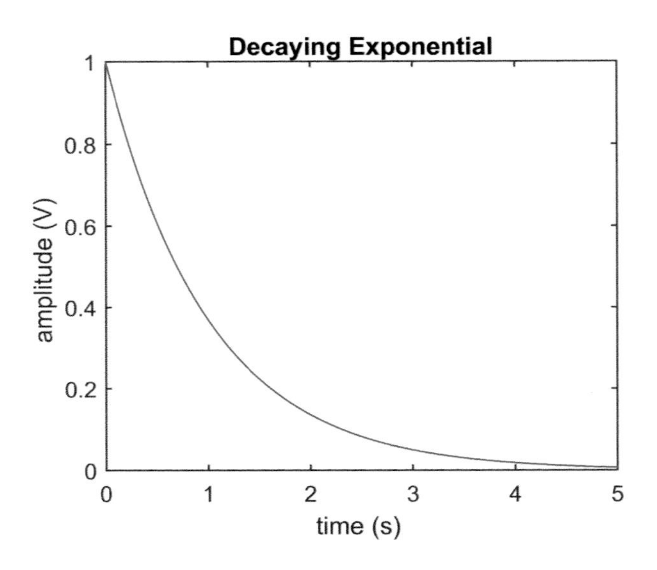


***Axis Limits***

MATLAB usually does a good job of choosing vertical and horizontal axis limits, but not always. For example, to make a right triangle, draw three lines connecting these four points (the last point is the same as the first to close the triangle): {0,0}, {1,0}, {0,1}, {0,0}. This corresponds to:

```
x = [0 1 0 0]; y=[0 0 1 0]; plot(x,y)
```

Here, the figure's axis limits are scaled to the limits of the figure, making the edges of the triangle difficult to distinguish from the axes themselves:

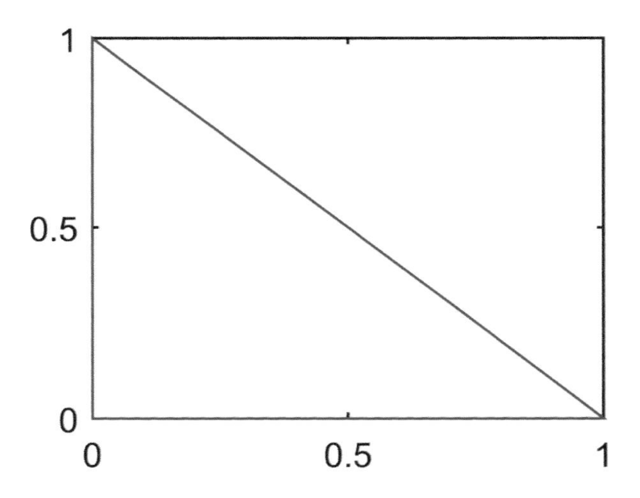

To choose your own axis limits, use the following command:

```
axis([xmin xmax ymin ymax])
```

For example, to clarify the plot in the previous figure, choose horizontal and vertical axis limits that range from -1 to 2:

```
axis([-1 2 -1 2])
```

This command will result in the much clearer figure below:

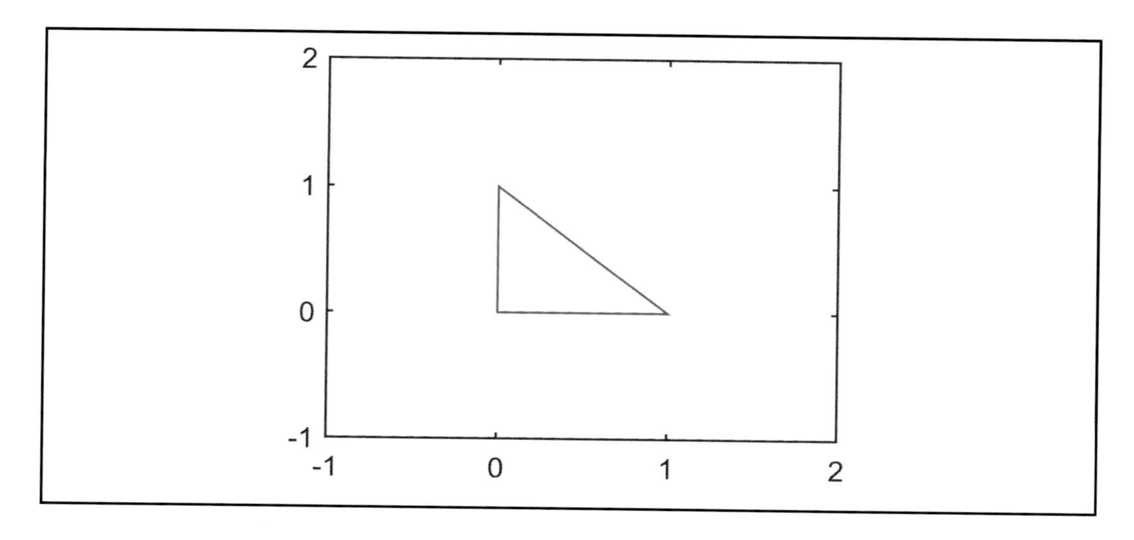


## *PRACTICE PROBLEMS*

3. Many waveforms in Electrical Engineering are derived from complex exponentials. Plot $y$ = real part of ($e^{jt\pi}$) for $0 \le t \le 5$. (Remember how to find the real part of a complex number?) Title the plot "Complex Exponential" and label the horizontal axis "time" and the vertical axis "Volts". ©®

*TECH TIP: OSCILLOSCOPES*

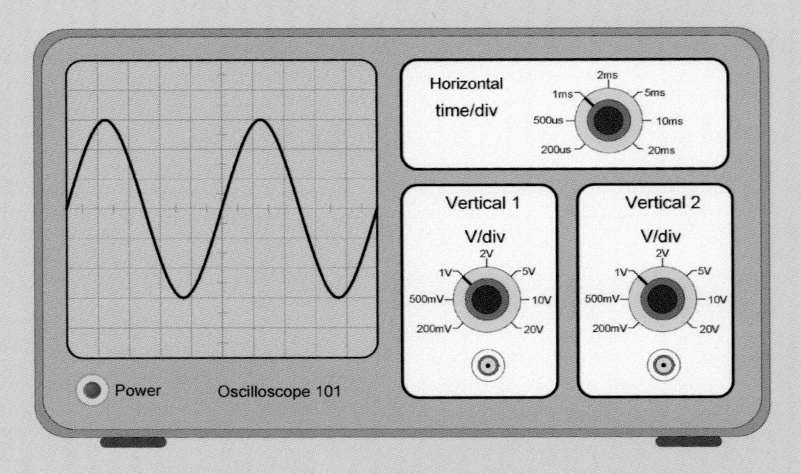


Electrical engineers often use oscilloscopes to graph time-varying voltages in circuits. Oscilloscopes are usually marked with a grid background. The user sets the scope so that each vertical division of the grid represents a certain number of volts (abbreviated V/div), and so that each horizontal division corresponds to a known amount of time (abbreviated time/div).

### *Grids*

You can add an oscilloscope-like grid to your plots with the `grid` command. After creating your basic plot, turn the grid on or off using these commands:

```
grid('on')
grid('off')
```

## PRACTICE PROBLEMS

4. Create an oscilloscope-like plot (that is, use a grid) of the waveform $x(t) = \cos(3t) - 2$ for $0 \le t \le 5$ seconds. Label the plot as "Oscilloscope trace," the horizontal axis as "time(s)," and the vertical axis as "Volts." ®

### *Logarithmic Axis Scaling*

Logarithmic axis scaling is useful when plotting values where the independent (horizontal) or dependent (vertical) variables have dense information for small values, but progressively spread-out larger values. For example, attempting to plot each of the 145 standard-value 5% resistor values from 1Ω to 1MΩ would be impossible on a standard, linear-scaled axis because the difference between the 0.1 Ω difference between the 1.0 Ω and 1.1 Ω would not be visible on a scale that extended to 1,000,000 Ω.

To fix this, the horizontal, vertical, or both axes of a plot may be logarithmically scaled, using the following commands, respectively:

```
semilogx(horizontal_data, vertical_data)
semilogy(horizontal_data, vertical_data)
loglog(horizontal_data, vertical_data)
```

These commands are used in place of the `plot()` command and have the exact same syntax. For example, to create a plot of log-scaled voltage data in vector `v` measured at corresponding times `t`, using a black line five times thicker than the MATLAB default,type:

```
semilogy(t,v,'k','linewidth',5)
```

Note that a logarithmically-scaled axis cannot contain a zero value, since $\log(0) = -\infty$.

Plots comparing the same data with and without logarithmic scaling of the horizontal axis are shown below. Notice how horizontal log scaling of this particular data set reveals important information lost in the linearly-scaled plot, even though the horizontal and vertical ranges of the axes are the same in both.

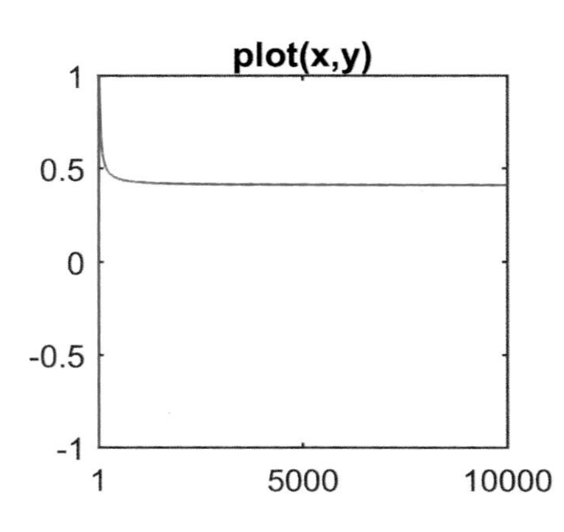


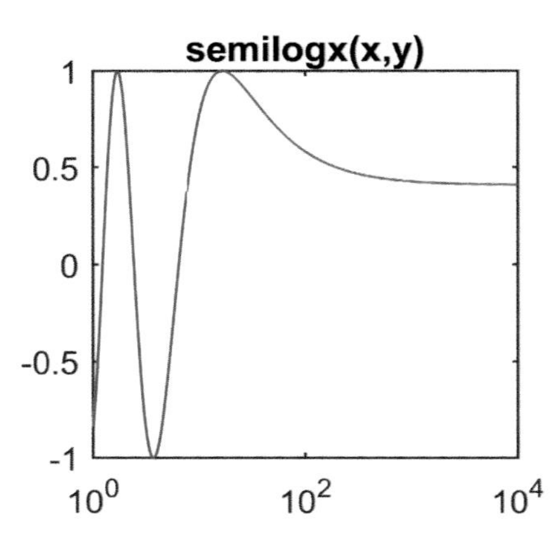

*TECH TIP: BODE PLOTS*

Bode plots are a specific type of plot used by electrical engineers to represent the amount of energy in a signal or passed by a filter as a function of frequency. The vertical scale is measured in decibels (abbreviated as dB), which are calculated as 20 $\log_{10}(x)$, where x is the signal energy. The horizontal scale is always plotted logarithmically. For example, to plot the frequency (Hz), voltage (V) pairs {1, 1}, {10, 1}, {100,0.1}, {1000,0.01} as a Bode plot in MATLAB:

```
f = [1 10 100 1000];
y = [1 1 0.1 0.01];
dB = 20*log10(y);
semilogx(f,dB)
title('Bode Plot')
xlabel('Frequency (Hz)')
ylabel('Amplitude (dB)')
```

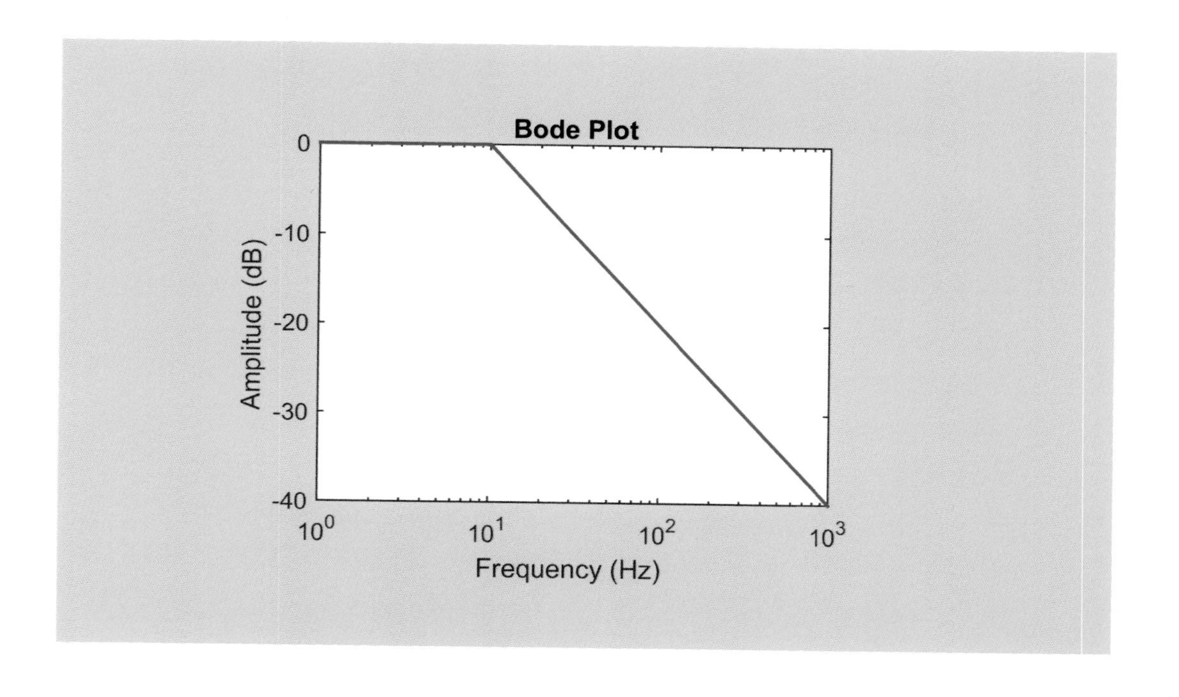


## PRACTICE PROBLEMS

5. Plot the following filter response using the above Bode-styling: $\left(y=\frac{1}{10000+f^2}\right)$. Use `logspace` to generate logarithmically-spaced points in the f vector from 1 Hz to 10,000 Hz. In other words, the Tech Tip gave the f and y vectors. Now, calculate them with `logspace` and the given equation. ® Hints:
   - If `logspace` returns infinite numbers, reread its section in Chapter 2.
   - Recall that `*`, `/`, and `^` work only between scalar and vectors. How do you modify them to work on respective elements between two vectors?
   - While the problem requires you to create your own horizontal and vertical vectors of data, the dB computation and the logarithmically-scaled axis plotting routine is the same as the example (and indeed, for any Bode plot).

*TECH TIP: LOW PASS FILTERS*

Low pass filters are a common type of electrical circuit that removes high frequencies and allows lower ones to pass through. The simplest type can be constructed from a single resistor and capacitor, as shown below on the left. Circuits analysis courses can teach you how to derive the equation shown below on the right, which completely describes the circuit's behavior. This equation is the circuit's **transfer function** $H(f)$, and evaluates to a complex number, as described in Chapter 2. To see what the circuit does to an input signal of frequency $f$, evaluate $H(f)$ at that frequency and take the magnitude of the resulting complex number. If $|H(f)| = 1$, it passes that frequency without change; if $|H(f)| > 1$, it amplifies the signal; and if $|H(f)| < 1$, it attenuates it.

R

$v_{in}(t)$ C $v_{out}(t)$

$$H(f) = \frac{1}{1 + j2\pi f RC}$$

**Simple low pass filter**

**Transfer function**

$H(f)$ is complex; it is the magnitude $|H(f)|$ that is commonly plotted against frequency $f$ and often as a Bode plot, plotting the $f$ along a logarithmically-scaled horizontal axis, and plotting the magnitude $|H(f)|$ in dB.

For example, plot the response of this circuit with $R = 1\ k\Omega$ and $C = 1.6\ \mu F$ from $1 \leq f \leq$ 10kHz with the following commands:

```
R = 1000;
C = 1.6e-6;                 defines the component values
f = logspace(0,4,100);      creates a log-spaced frequency
                            vector from 1 to 10,000 Hz
H = 1./(1+j*2*pi*f*R*C);    calculates the H vector
dB = 20*log10(abs(H));      calculates the magnitude
                            of H in dB
semilogx(f,dB)
title('Lowpass filter')
xlabel('Frequency (Hz)')
ylabel('Amplitude (dB)')
grid('on')
```

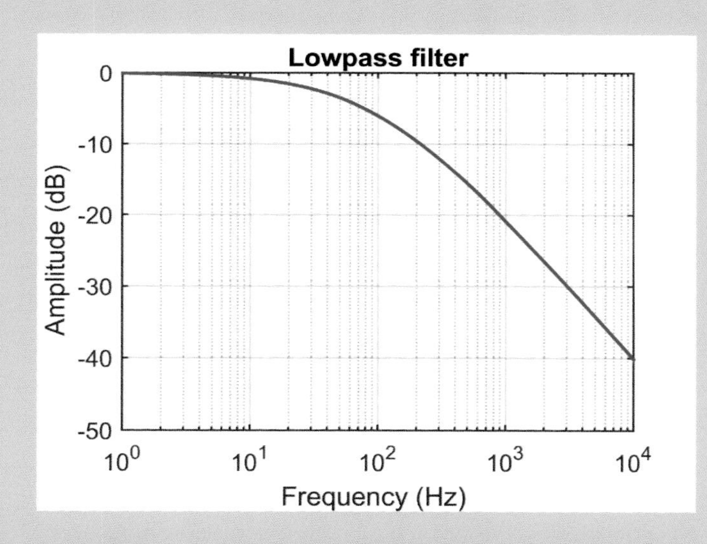


A lowpass filter response is shown above. It starts to cut frequencies above about 100 Hz.

## PRACTICE PROBLEMS

6. Re-do the Tech Tip example above using the values R = 5kΩ and C = 3.2μF. Plot |H(f)| in dB for 30 frequencies that logarithmically span from 0.1Hz to 10kHz. ®

## LINE PLOTS WITH MULTIPLE LINES

Just as oscilloscopes often have two channels of input data, so engineers often graph more than one set of data on the same axes. Both datasets {t1, y1} and {t2, y2} can be plotted on the same axes with the following command:

```
plot(t1,y1,t2,y2)
```

The following commands plot $y1 = \sin(t)$ and $y2 = \cos(t)$ in the same axes for $0 \le t \le 6$:

```
t = linspace(0,6,1000);
y1 = sin(t);
y2 = cos(t);
plot(t,y1,t,y2)
```

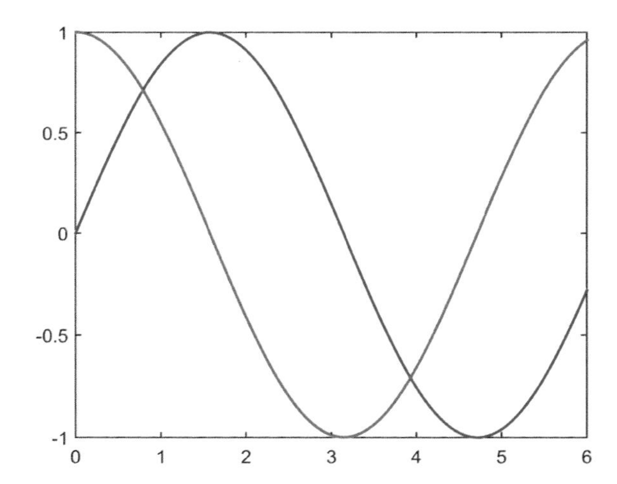

Note that all of the options introduced previously work as expected. The following commands will make the first line green and the second black and dashed, with each plotted at twice the default MATLAB width (see the corresponding figure below):

```
plot(t,y1,'g',t,y2,'k-','linewidth',2)
```

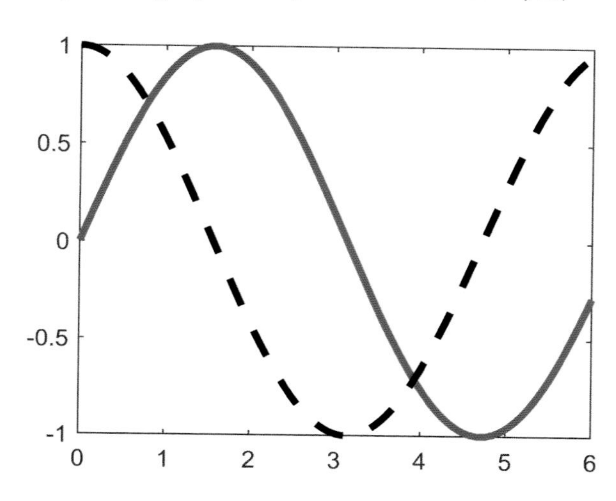


There is no limit to the number of lines plotted on one set of axes; for example, with more calculated values of *y*3, *y*4, and so on, one could use a command like:

```
plot(t1,y1,t2,y2,t3,y3,t4,y4,…)
```

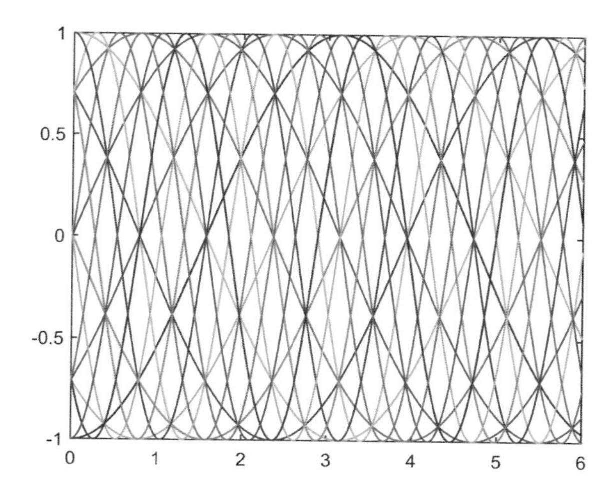

### ***Adding Legends***

A figure legend enables a viewer to tell the difference between different lines plotted on the same set of axes. To add a legend to a set of two lines, use the following syntax:

```
legend('name1', 'name2')
```

To add a legend to the sin/cos plot our earlier example, type:

```
legend('sin(t)', 'cos(t)')
```

The legend may be dragged with the mouse to avoid covering the data.

Legends are especially useful when printing on black and white printers. In that case, denote the difference between plots by using solid and dashed lines.

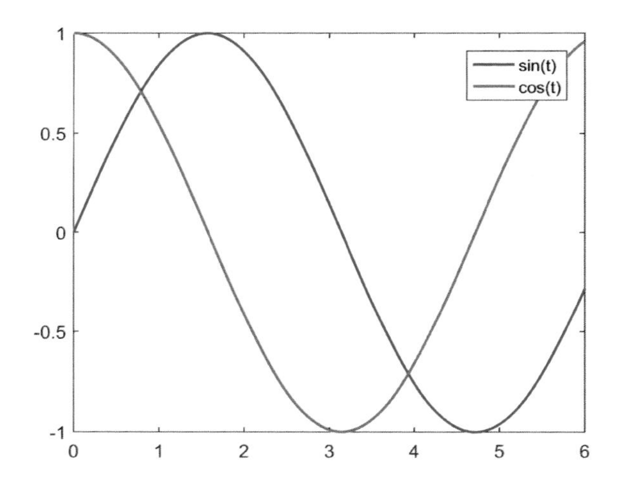


*PRACTICE PROBLEMS*

7. A circuit is found to have a response $v_1(t) = 2e^{-t}$, for $0 \leq t \leq 1$. Adding a capacitor introduces oscillations, making the new response $v_2(t) = 2e^{-t}\cos(20t)$, over the same time region. Plot both $v_1(t)$ and $v_2(t)$ on the same set of axes, label the axes, and create a legend. Use as many points as needed to generate a smooth plot. ®

## SCATTER PLOTS

Scatter plots use similar data to line plots, but instead of joining each {x,y} pair of data with lines, scatter plots mark the points themselves with markers. They are most appropriate when graphing discrete data values, where the linear interpolation between data points that is implied by connecting them with a line may not be appropriate. To create scatter plots, use the same syntax as for a plot, but use one of the symbols in the table below instead of a line style symbol:

| Symbol | Abbreviation |
|---|---|
| • | '.' |
| o | 'o' |
| × | 'x' |
| + | '+' |
| * | '*' |
| Δ | '^' |
| □ | 's' |

To create a scatter plot for measured resistances r = [57 65 63 59 65] of five different heating elements x = [1 2 3 4 5], use the following code:

```
x = 1:5;
r = [57 65 63 59 65];
plot(x,r,'o')
```

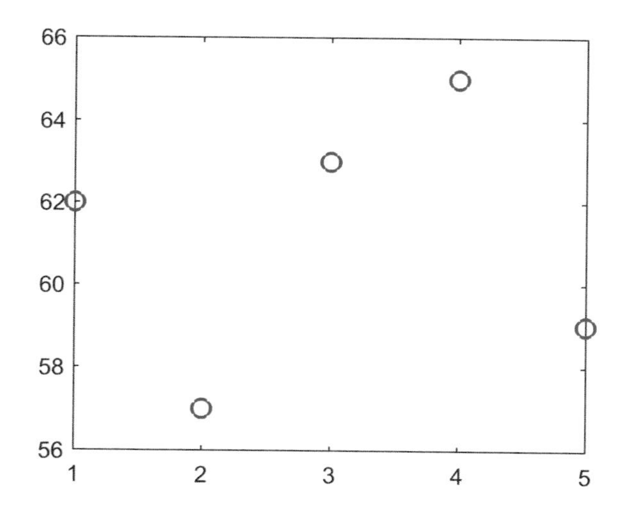

### ***Decorating Scatter Plots***

The same techniques learned earlier with `plot()` to specify marker color, title, axis labeling, and grid also work for scatter plots (now, color specifies the marker color rather than the line color). You can also specify the marker size with the `'markersize'` keyword. For example, the plot from the previous page could be changed to the plot below with the following command:

```
plot(x,r,'^','color',[1 0.5 0],'markersize',25)
```

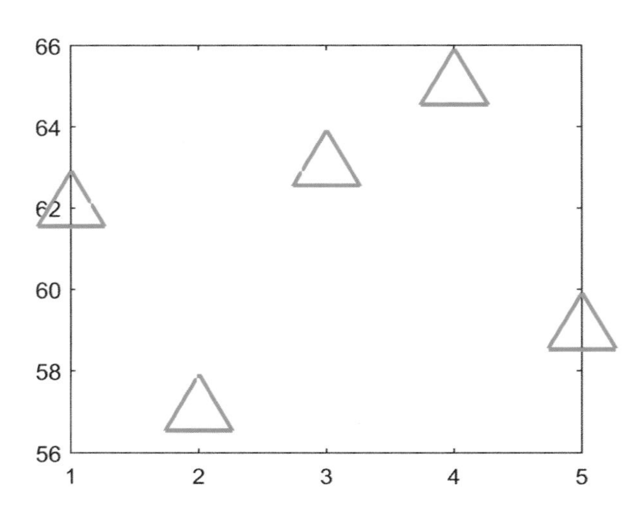


PRACTICE PROBLEMS

8. Five 10kΩ resistors are pulled from a drawer and measured. Their values are found to be 9.6kΩ, 10.2kΩ, 10.4kΩ, 9.9kΩ and 10.0kΩ. Scatter plot these values using circles as markers. For the horizontal axis, use the measurement numbers 1, 2, 3, 4, 5. Label the axis and title the plot. ®

### *PRO TIP: PROFESSIONAL LICENSURE (PE)*

Just as doctors and lawyers are licensed to practice medicine and law, engineers can be licensed as well. Licensed engineers are called "Professional Engineers," and they add the letters P.E. after their name, like John Doe, P.E. Unlike law and medicine, most electrical engineering careers do not require licensure (however, such requirements are becoming increasingly more common). Careers in which licensure is especially helpful include:

- Architecture and Building: Only P.E.s may certify certain types of engineering plans.
- Consulting: Only P.E.s may engage in private engineering consulting, and engineering consultant firms are required to maintain a certain minimum ratio of licensed engineers on their staff.
- Government: Being a P.E. is a common requirement for higher-level state and federal engineering jobs
- Energy: Most power distribution companies seek P.E.s to help them certify power station plans.
- Teaching: Some states require engineering departments to ensure a certain percentage of their engineering faculty are licensed.

Licensing requirements vary by state, but a typical path includes graduating from a four-year engineering program, passing the Fundamentals of Engineering (F.E.) exam, working for four years under a P.E., and then passing the P.E. exam.

## PLOTTING SCRIPTS

Students often find themselves copying many previously-typed lines of commands when modifying complicated plots. There is a way to do this much more efficiently by creating a MATLAB script.

1. Navigate into your personal directory using the MATLAB folder window and directory toolbar, as highlighted below:

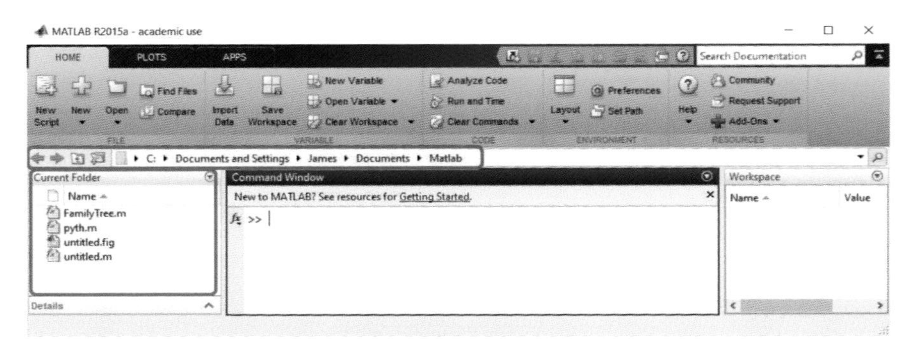


2. From the MATLAB command window type `edit`. This will spawn a separate text editing window. In the text editing window, type the commands shown below (in this case, to plot a sinc waveform):

```
t = linspace(-10,10,100);
y1 = sin(t)./t;
y2 = zeros(size(y1));
plot(t,y1,'b-',t,y2,'k-')
```

3. Save the file as "figure1.m". All MATLAB code files must end with a .m suffix. The edit window should now look like the figure below:

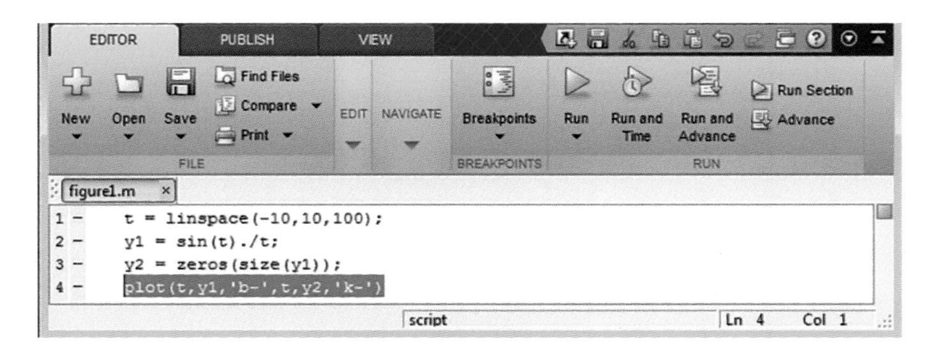

4. To run the code, a *script*, go back to the main MATLAB command window and type the name of the file without the suffix. In this case, to run the script in figure1.m and generate the figure below, type `figure1` in the command window.

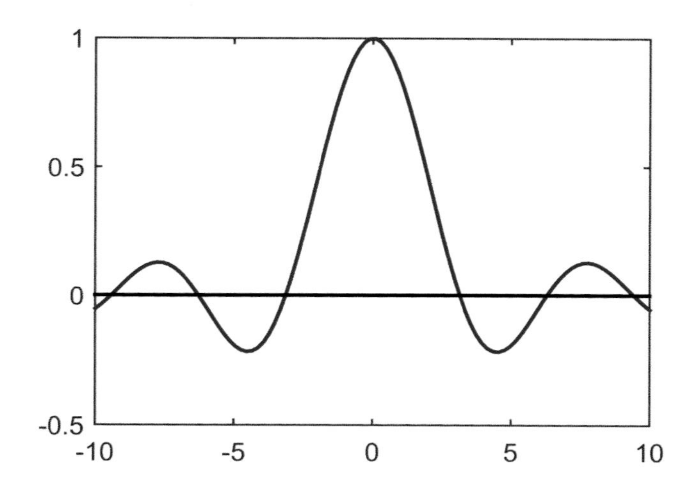


The extra work to create a script to generate complex plots is often justified by the ease with which they can be modified. In our current example, you can make the plot lines twice as thick just by changing the last line to add `'linewidth',2` and re-running the script.

### PRACTICE PROBLEMS

9. Create a script to build the sinc waveform given in the previous example, and then modify the script to make the linewidth five times as thick as the default. Run the script (do not cut and paste the commands into MATLAB). Show both the script and the result. ©®

## LAYERING PLOT COMMANDS USING hold()

One of the most powerful abilities of `plot()` is its ability to layer multiple line plots on the same axis. There are two different ways to accomplish this:

1. Use the `plot(x1,y1,x2,y2)` syntax earlier introduced
2. Use `plot(x1,y1),hold('on'),plot(x2,y2),hold('off')`

Let's begin by describing the weakness of the first method. You know how to plot two sets of data {x1, y1} and {x2, y2} with two lines using:

```
plot(x1,y1,x2,y2)
```

You also know how to format them individually—for instance, making the first line red and solid, and the second line black and dotted. Recall the commands must be grouped with the data, as in the example below:

```
          first line      second line
plot(x1,y1,'r-',x2,y2,'k:')
```

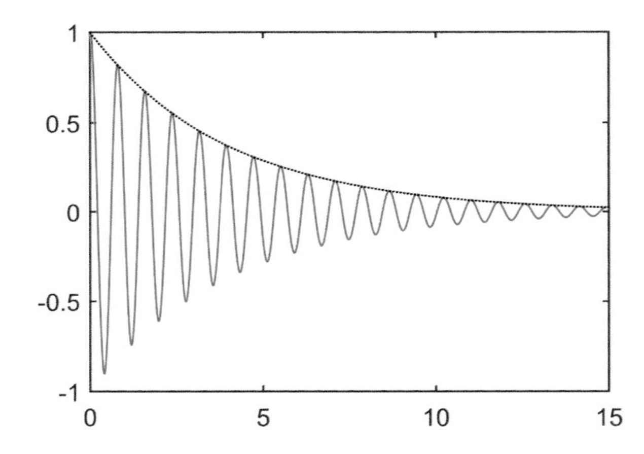


The second black dotted line here is hard to see because it is small. How can you make the first line thick and the second line thin? Two-part commands like `'linewidth',5` and `'color', [1 0 1]` will not work because they must appear at the very end of the `plot()` command, and so they apply to all the line plots within the `plot()` command, as shown below:

first line second line whole-plot line

```
plot(x1,y1,'r-',x2,y2,'k:','linewidth',5)
```

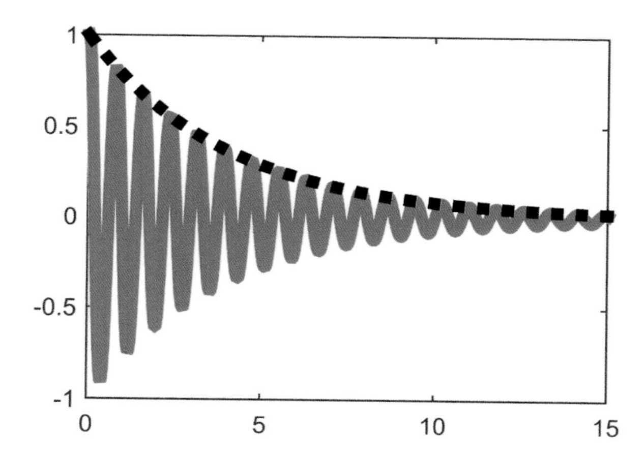


To obtain the desired effect of just the second (black) line being thicker, one must use the `hold()` command to layer two different `plot()` commands on top of each other. To do this, plot only the first line, and then turn hold on using `hold('on')`, as follows:

```
plot(x1,y1,'r-')
hold('on')
```

Then plot the second thick black line with the linewidth modifier:

```
plot(x2,y2,'k:','linewidth',5)
```

Do not forget to turn hold off, or the next plot you create will continue to layer on top of the current plot:

```
hold('off')
```

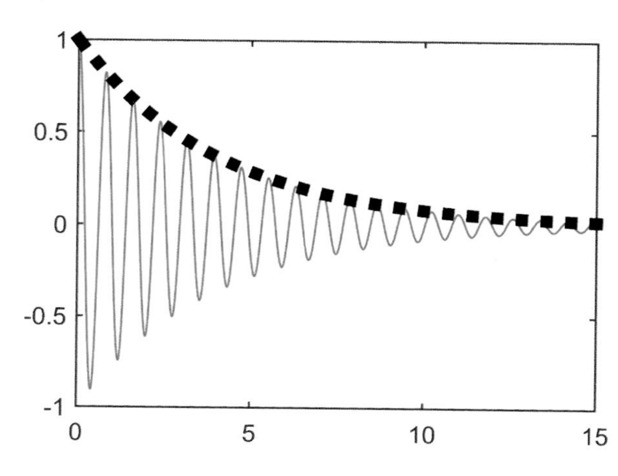

## PRACTICE PROBLEMS

10. Create a plot of $h(t) = e^{-t}$ from $0 \leq t \leq 5$ using a black line 15 times thicker than the default. Using `hold()`, plot the default thin-sized line of $y(t) = e^{-t} + 0.04 \cos(15t)$ over it in white (that is `'w'`). Remember to turn hold off at the end. ®

## BAR PLOTS

A bar plot is a convenient way to visualize large quantities of raw data grouped into bins. For example, if a large class completes a lab and measures the following voltages across a capacitor:

| $V_c$ | 3.3 | 3.4 | 3.5 | 3.6 | 3.7 |
|---|---|---|---|---|---|
| Number of observations | 3 | 12 | 34 | 25 | 9 |

Then to graph this relationship, use the same syntax as the plot command, but use `bar()`. As always, the horizontal data go first, followed by the vertical data:

```
v = 3.3:0.1:3.7;
n = [3 12 34 25 9];
bar(v,n)
```

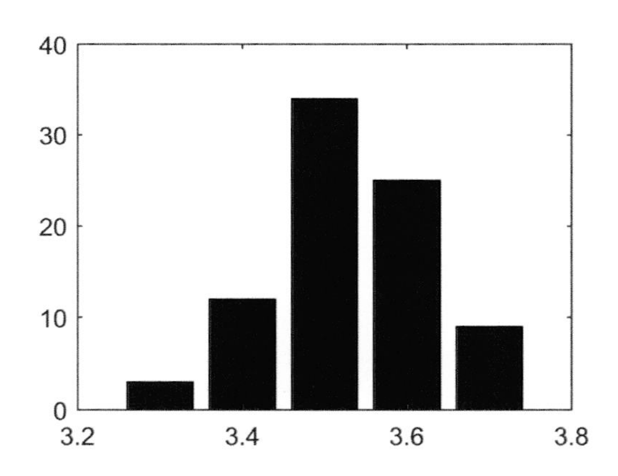

The same techniques used to adorn line plots also work with bar plots. For example, the previous plot could have a title and axis labels applied using `title()`, `xlabel()`, and `ylabel()`, or the `axis()` command could be used to change the axis scaling to eliminate the white space at the right side of the plot.

Bar plots can be colored and also grouped into different series as shown below; type `help bar` for more information.

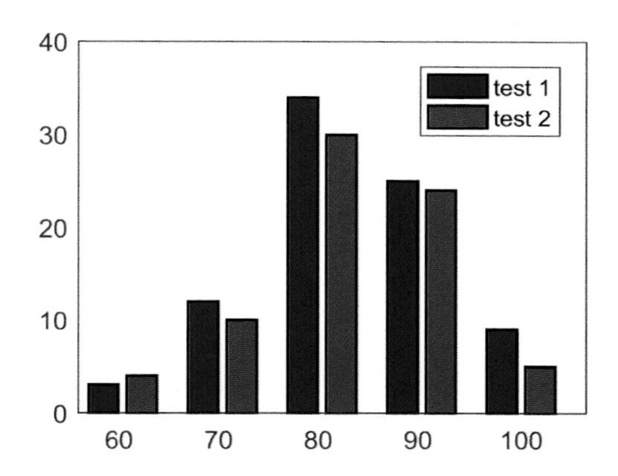


*PRACTICE PROBLEMS*

11. Digital logic gates have a (usually undesired) delay between when an input signal changes and when the gate responds, called propagation delay $t_p$. The following propagation delays are tested for the common 74LS04 chip, a NOT gate.

| $t_p$ | 2.5ns | 2.75ns | 3ns | 3.25ns | 3.5ns |
|---|---|---|---|---|---|
| # of chips | 2 | 12 | 27 | 16 | 3 |

Create a bar plot displaying the above values. Label the axes and give it a title. Hint: Rather than converting very small numbers like 2.5 ns to 0.0000000025 s, leave it as 2.5 and label the appropriate axis as "time (ns)". ®

## SUBPLOT

You have learned how to plot multiple items on the same set of axes using `hold()` or `plot(x1,y1,x2,y2)`. Sometimes you may want to group multiple separate axes in the same plot as in the figure below, which shows the frequency response of a Butterworth lowpass filter plotted against various combinations of axis scaling. Note that neither `hold` nor `plot` can put these on the same set of axes, since it is the same data plotted using different axis scalings.

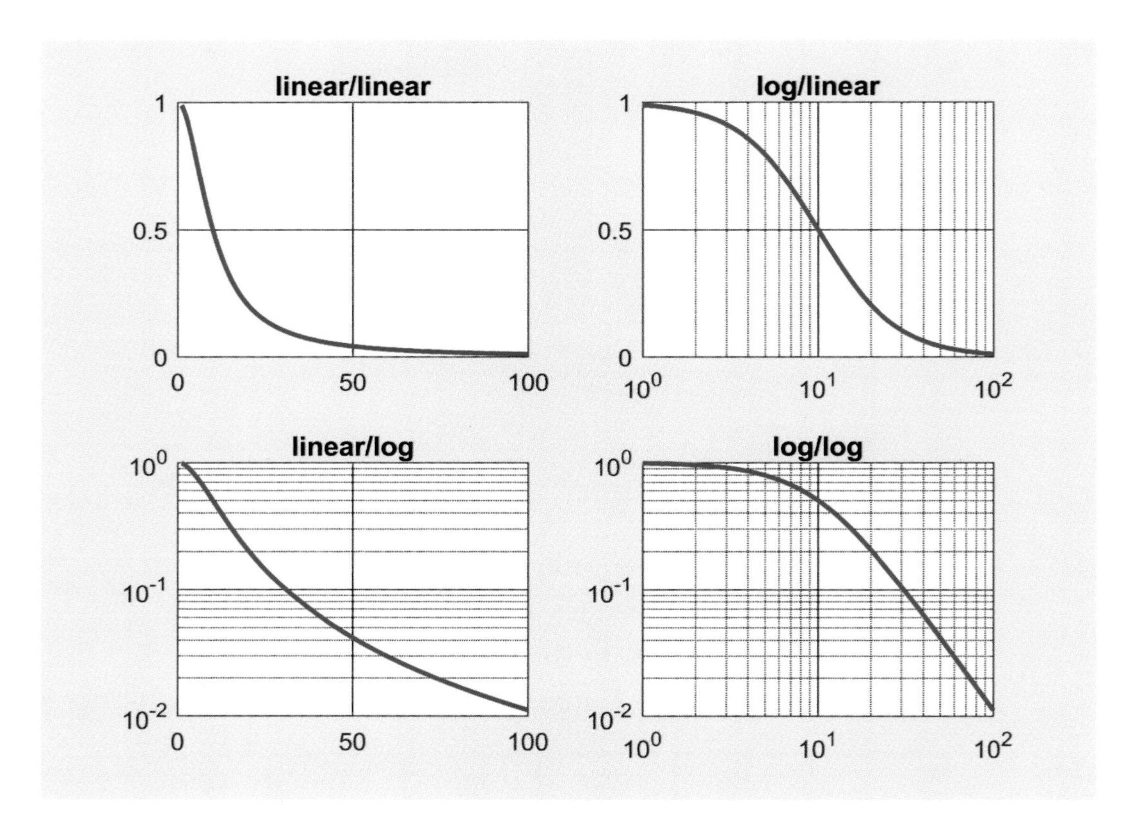


Instead, use the `subplot()` command. This command divides the master plot into rows and columns of sub-plots (2 rows and 2 columns in the above example), and sets the "active" sub-plot, so that the next `plot()` command will draw into it. To create a set of *r* rows by *c* columns of sub-plots, and to set the active sub-plot to *p*, use the command:

```
subplot(r,c,p)
```

The sub-plots are numbered in the same way that we read in English, from left to right, then top to bottom, as shown below:

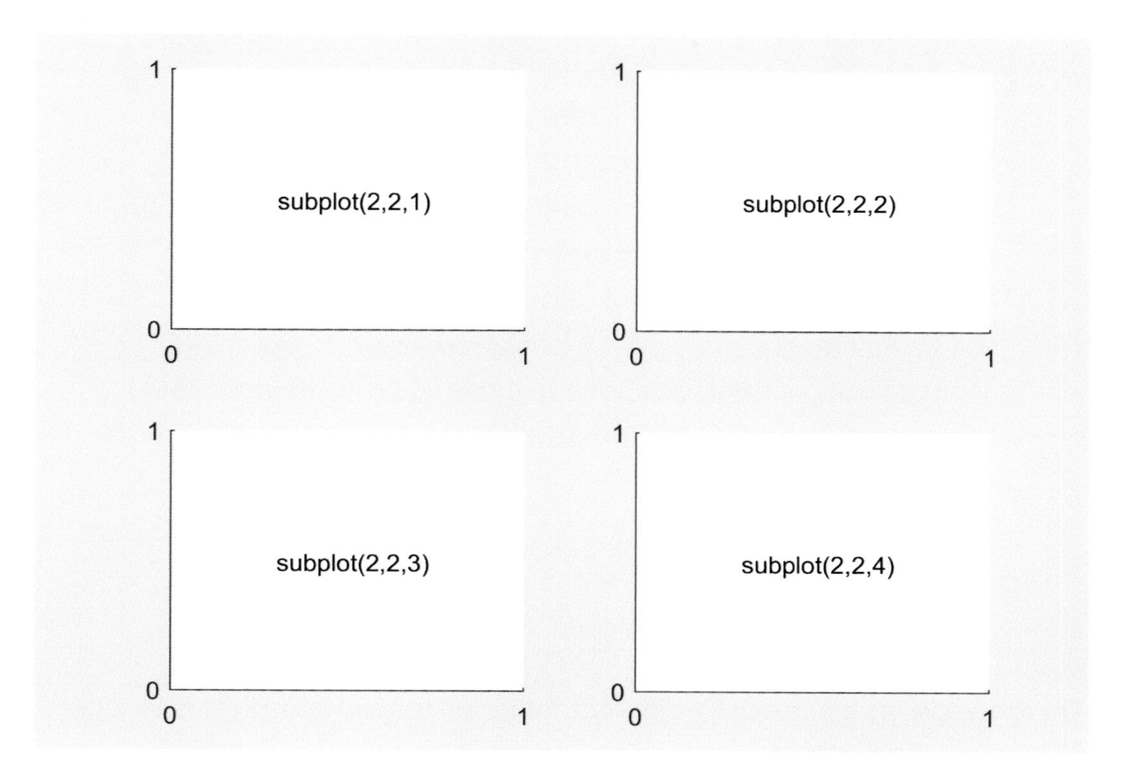


As an example, the following commands create a set of two long narrow plots, one over the other, the top one graphing $v_1(t) = \sin(t)$, and the bottom graphing $v_2(t)$ = a triangle wave, for $0 \le t \le 4\pi$:

```
t1 = linspace(0,4*pi,1000);
v1 = sin(t1);
t2 = linspace(0,4*pi,9);
v2 = [0 1 0 -1 0 1 0 -1 0];

subplot(2,1,1)
plot(t1,v1)
title('sin(t)')

subplot(2,1,2)
plot(t2,v2)
title('triangle wave')
```

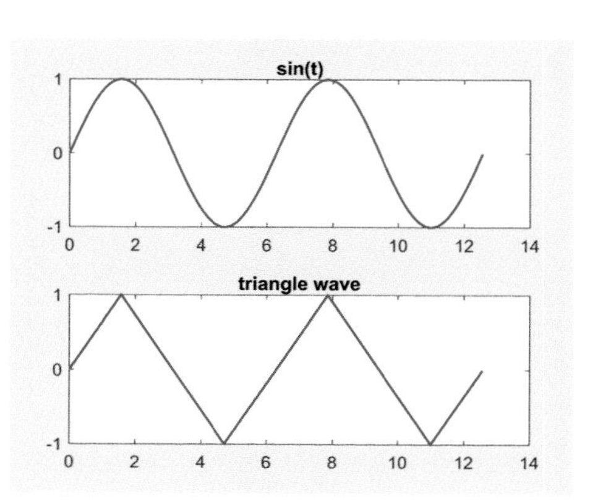

## PRACTICE PROBLEMS

12. A circuit with inductors, capacitors and resistors "rings" (oscillates) when driven by a step in voltage. Both voltage and current oscillate with the same frequency but with different phases, such as $v(t) = e^{-t/4}\cos(4t)$ and $i(t)=0.01\ e^{-t/4}\cos(4t+\pi/2)$. Plot for $0 \le t \le 15$. Plot these in two subplots, one on top of the other like in the above example. ®
13. Using the `plot(x,y1,x,y2)` command, plot the above data using a single axis. Which of these two methods is a better way to visualize the similarities between v($t$) and $i(t)$? ®

# ADVANCED PLOT DECORATION

There are two fundamentally different methods to gain complete control over all aspects of plots:

- the interactive plot editor
- handle graphics

### *Interactive Plot Editor*

To enter the interactive plot editor, first create a plot, and then from the plot's figure window, menu choose Tools → Edit Plot.

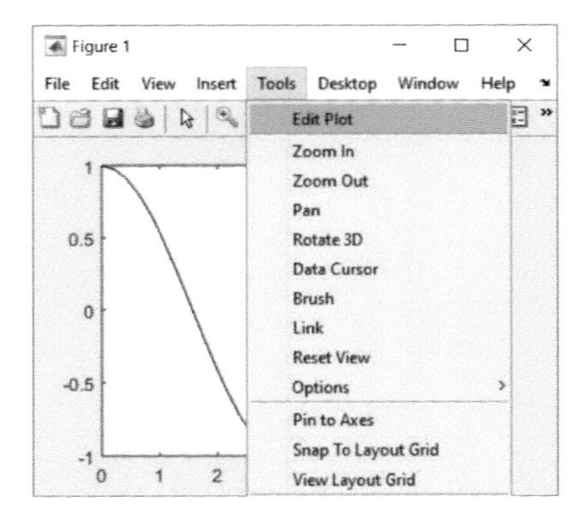

Square selection boxes appear in this mode. There are three selectable things, as shown below: the Figure Window, the Axis Window, and the Data Set.

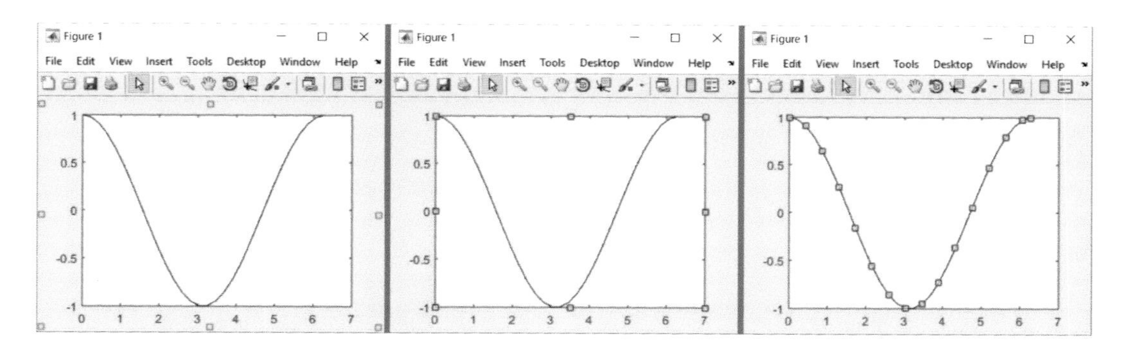

The Figure Window (left), the Axis Window (middle), and Data Set (right) selected when in Edit Plot mode

Once the desired item is selected, double-click to enter the editor. The editor is context-sensitive and will offer choices dependent on the currently-selected item. Because it is graphical and interactive, the interface is very intuitive to use. This is the primary benefit to using this method: it is fast.

### ***Handle Graphics***

An entirely different method to fully customize your graphics is to use handle graphics. This method uses keyword/value pairs, like `'linewidth', 5`. The difficulty with this method is that there are many different keyword pairs, and they are accessed through handle graphics, which means you will need to use the new commands `get()` to learn which keywords are available, and `set()` to change them. If you need to print an idea quickly, you might find the interactive plot editor preferable. However, in practice, this is rare; if you need to adjust details of the plot, it is often because you plan to use the plot in a document or presentation (which you may need to revise at a later date). All changes made using the interactive plot editor are lost, whereas with handle graphics, one can build a script file to build and modify each figure, and then recreate it later, thus preserving the document's editability.

### *Using get() to Get the Keyword Names of Settable Properties*

To use handle graphics, it is necessary to understand how MATLAB stores graphics information. All information about the plot, from data, to line style, to axis tick labels, is stored inside one of three elements: the figure window, the axis, or the line plot. These are the same three elements that are selectable using the interactive editing mode. To see the options for each element, use `get()`

#### *Figure window*

`get(gcf)` lists figure window properties. There are about sixty of these properties, including:

`'position', [x,y,width,height]` sets the size and location of the figure window relative to the desktop

`'color', [r,g,b]` sets the color, in [red green blue] coordinates, for the background of the figure window.

#### *Axis*

`get(gca)` lists all axis properties (there are over 120), including:

`'position', [x,y,width,height]` sets the size and location of the axis window relative to the figure window;

`'xtick', 1:10` sets the numeric position of the tick marks along the x axis;

`'fontSize', 12` sets the font size in points of the text labeling the axis

`'visible', 'off'` makes the axes invisible, but keeps visible the data that they display

#### *Data*

`h = get(gca,'children');get(h)` lists the properties of the plotted data. The number of these properties depends on the type of plot, but data for a line plot has about thirty-six of them, including:

`'markerfacecolor', [r,g,b]` sets the fill color for the data markers;

`'markeredgecolor', [r,g,b]` sets the color for the outside border of the data markers

`'linecolor', [r,g,b]` sets the fill color for the data markers.

*DIGGING DEEPER*

With well over 200 different keywords that govern even simple graphics objects, this subject could be a book in itself. It is even possible to attach actions to graphics objects, so that clicking them, or having the mouse pass over them, executes other MATLAB code called callback functions. If you are interested in this, finish the Programming chapters first, and then consult the MATLAB help for GUIDE, which lets you write your own interactive GUI programs.

### *Using set() to Get the Keyword Names of Settable Properties*

To set the properties, use the `set()` command instead of the `get()` command, and use the two-part keyword/value for the property you wish to set, e.g.:

**Figure** (keyword: `'position'`, value: `[50,50,500,100]`)

`set(gcf,'position',[50,50,500,100])` makes the figure window wide (500 pixels) and short (100 pixels), and places it in the lower left-hand corner of the desktop.

**Axis** (keyword: `'xtick'`, value: `[0 1]`)

`set(gca,'xtick',[0 1])` changes the horizontal tick locations to 0 and 1

**Data**

```
h = get(gca, 'children');
```

(keyword: `'markerfacecolor'`, value: `[1,0,1]`)

`set(h,'markerfacecolor',[1,0,1])` fills the markers with pink

### PRACTICE PROBLEMS

14. Create a scatter plot of the data x=[.5 1.5 1], y=[.5 .5 1.4], using black circles of size 20. Hint: after you create the data, you can plot this with a single command. Use the interactive plot editor to fill them with green. ®
15. Create a script file that does the same as the above using the handle graphics set command. Hint: Changing the marker face color is a property of the data. ©

## TEXT ANNOTATIONS

To create text at an arbitrary position on a plot, use the following command:

```
text(x,y,'text string')
```

The example below illustrates this technique, and shows the resulting plot:

```
t = linspace(0, 5, 1000);
y = t/2 .* cos(t);
plot(t,y)
text(0.5,0.5,'Local maxima')
text(3,-1.8,'Global minima')
```

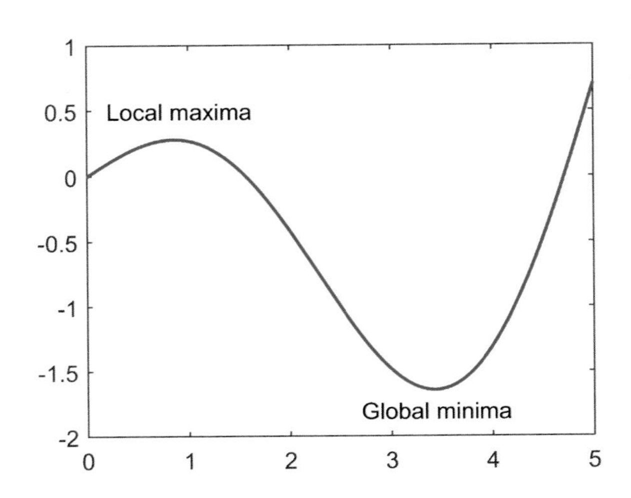

### *TECH TIP: GREEK SYMBOLS IN ECE*

Electrical engineers commonly use 19 of the 24 letters in the Greek alphabet, and it is important to be able to recognize and pronounce the names of them. The table below pairs each symbol with its application in ECE:

| | | |
|---|---|---|
| α | alpha | exponential decay rate |
| β | beta | transistor current gain |
| Γ | Gamma | transmission reflection |
| Δ | Delta | change |
| δ | delta | impulse function |
| η | eta | efficiency ratio |
| θ | theta | angle |
| λ | lambda | exponential decay rate |
| μ | mu (micro) | $10^{-6}$ |
| ξ | xi | damping ratio |
| Π | Pi | product |
| π | pi | 3.14159... |
| Σ | Sigma | sum |
| σ | sigma | conductivity, real part of a complex number |
| τ | tau | time constant |
| Φ | Phi | magnetic flux |
| φ | phi | angle (similar use as θ) |
| Ω | Omega (Ohm) | unit of resistance |
| ω | omega | radian frequency |

## ADVANCED TEXT FORMATTING

Electrical Engineering plots often require exponents, subscripts, or Greek symbols, and thus, MATLAB provides a way for `title()`, `xlabel()`, `ylabel()`, `legend()`, and `text()` to display them.

### ***Greek Symbols***

Greek symbols are very common in Electrical Engineering, as described in the previous Tech Tip, and they can be embedded using the backslash character \, followed by the name of the Greek symbol. The following command labels the horizontal axis with "Resistance in MΩ":

```
xlabel('Resistance in M\Omega')
```

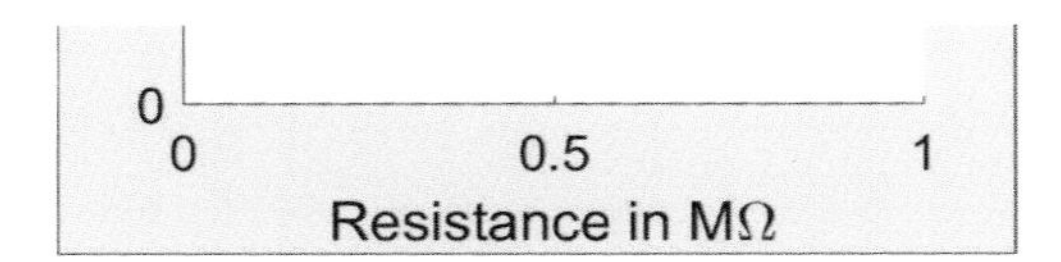


### ***Super and Subscripts***

Superscripts, like powers, such as $t^2$, and subscripts, like $V_1$, are easy to create in MATLAB text if the superscript or subscript is just a single character: use ^ or _ respectively. The following command creates the title $V_1(t) = f(t^2)$:

```
title('V_1(t) = f(t^2)')
```

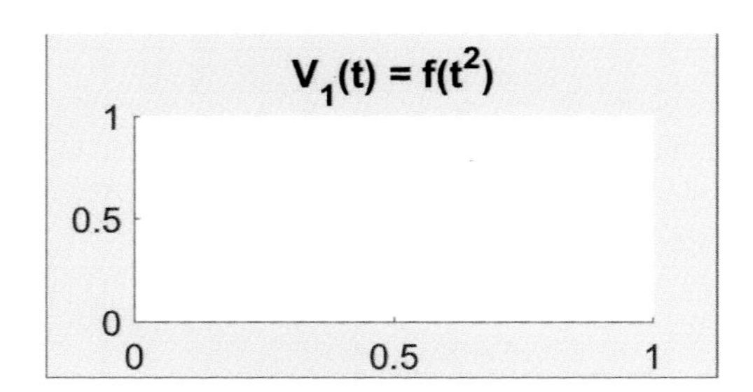


To group multiple characters in the superscript or subscript, surround them with curly braces (i.e. {}). The following command creates the text annotation $f(t) = e^{j\omega t}$ (notice how it also combines Greek symbols):

```
text(0.25,0.5, 'e^{j\omegat}')
```

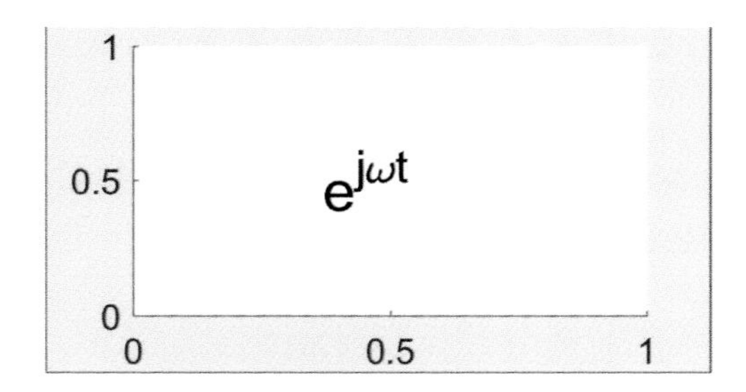

## PRACTICE PROBLEMS

16. Graph $v_1(t) = 2\,e^{-\lambda t}$, $0 \leq t \leq 0.5$, where $\lambda = 4$; label the vertical axis "$v_1$ (V)" and give it the title "$2e^{-\lambda t}$, $\lambda = 4$". ®

## THREE-DIMENSIONAL MATLAB PLOTS

Although the two-dimensional line, scatter, and bar plots described so far are the most common types of graphics used by electrical engineers, three-dimensional plots can show how one dependent variable changes as a function of two independent variables. For example, the sinc function discussed earlier in this chapter also exists as a function of two dependent variables, and it is used in image processing. This function is described by the equation:

$$z(x,y) = \frac{\sin\left(\sqrt{x^2+y^2}\right)}{\sqrt{x^2+y^2}}$$

To graph this between $-15 \leq x \leq 15$ and $-15 \leq y \leq 15$, first create $x$ and $y$ matrices using `meshgrid()`, which can be thought of to be a 3D counterpart of `linspace()`. Meshgrid's first argument is the vector of x locations, and its second is the vector of $y$ locations, as follows:

```
[x,y] = meshgrid(-15:15,-15:15);
```

Then define the function to plot:

```
z = sin(sqrt(x.^2+y.^2))./sqrt(x.^2+y.^2);
```

Last, create the 3D plot using `surf()`:

```
surf(x,y,z)
```

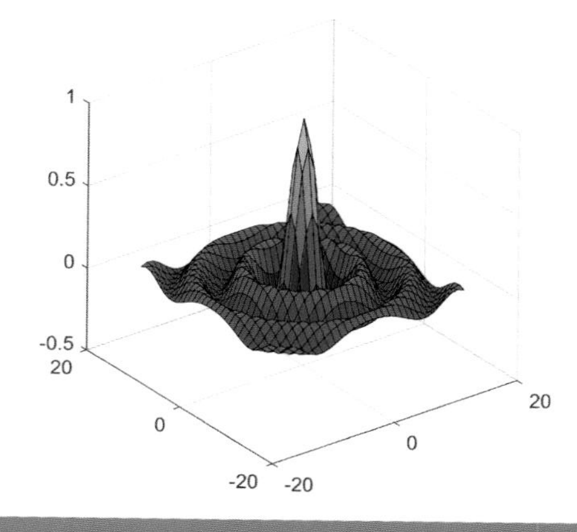

To get a smoother-looking plot, add a `shading()` command, which will turn off the black edges and make the shading more gradual:

```
shading('interp')
```

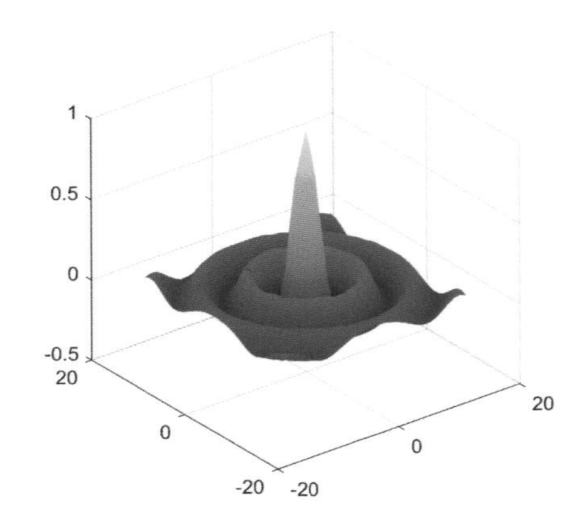


Solid surfaces are created by the `surf()` command. To create a wireframe, use `mesh()` instead of `surf()`, as shown below:

```
[x,y] = meshgrid(-15:15,-15:15);
z = sin(sqrt(x.^2+y.^2))./sqrt(x.^2+y.^2);
mesh(x,y,z)
```

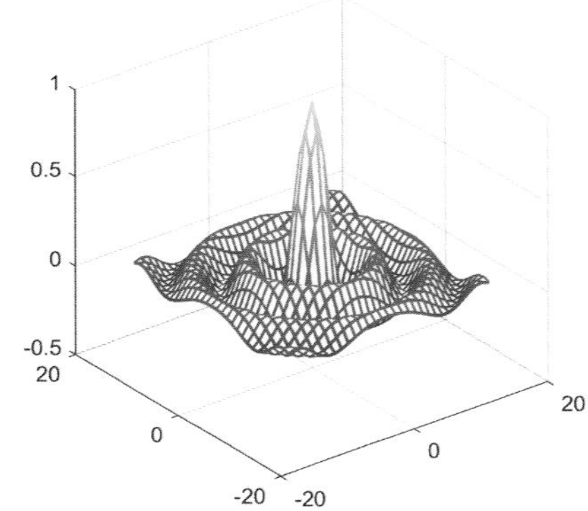


## *PRACTICE PROBLEMS*

17. Create a surface plot of the following function using a grid spanning both *x* and *y* from -10 to 10. ®

$$z(x,y)=\sin\left(\frac{x}{3}\right)\cos\left(\frac{y}{3}\right)$$

## PRO TIP: IEEE CODE OF ETHICS

To follow the IEEE Code of Ethics, electrical engineers agree to:

1. accept responsibility in making decisions consistent with the welfare of the public, and to disclose promptly factors that might endanger the public or the environment;
2. avoid real or perceived conflicts of interest where possible, and to disclose them to affected parties when they do exist;
3. be honest and realistic in stating claims or estimates based on available data;
4. reject bribery in all its forms;
5. improve the understanding of technology and potential consequences;
6. improve our technical competence and undertake technological tasks for others only if qualified by training or experience, or after full disclosure of pertinent limitations;
7. seek and offer honest criticism of technical work, to correct errors, and to credit the contributions of others;
8. treat fairly all persons and not to engage in acts of discrimination based on race, religion, gender, disability, age, national origin, sexual orientation, gender identity or expression;
9. avoid injuring others, property, reputation, or employment by false or malicious action;
10. assist colleagues with professional development and support them in following this Code.

*See the unabridged text at: https://www.ieee.org/about/ethics.html*

## COMMAND REVIEW

### *Line Plots*

`plot(x,y)` plots the x vector against the y vector

`plot(x,y,'r-')` red, straight lines. g, b, k = green, blue, black. -, --, : = solid, dashed, dotted

`plot(x,y,'r-', 'linewidth',5)` plots using a line width 5x the default

`plot(x,y, 'color',[r g b])` plots using a line with color specified by r, g, b values from 0 to 1

`plot(x,y,'b-',x2,y2,'r:')` plots two lines on the same axes (first a blue solid line, then red dotted)

`semilogx(x,y)` plots using a logarithmic horizontal axis

`semilogy(x,y)` plots using a logarithmic vertical axis

`loglog(x,y)` plots using logarithmic horizontal and vertical axes

### *Scatter Plots*

`plot(x,y,'k.')` plots points with dots. Use .,o,x for dot, circle, x markers

`plot(x,y,'o', 'markersize',10)` plots using a marker 10x default size

### *Bar Plots*

`bar(x,y,'b')` creates a blue bar plot. Other colors same as `plot()`.

### *Options for Line, Scatter, and Bar Plots*

`axis([xmin xmax ymin ymax])` sets axis limits on any plot, overriding the default

`grid('on'), grid('off')` changes the grid visibility

`legend('label1', 'label2')` creates a legend to label a plot with multiple sets of lines or markers

`hold('on'), hold('off')` causes next plot command to overlay data on top of previous

`subplot(r,c,n)` divides plot into r rows and c columns of different axes. The next plot will go into the nth one.

***Text on Line, Scatter, and Bar Plots***

`title('My title')` places a title at the top of a plot window

`xlabel('x axis')` places the text 'x axis' below the horizontal axis of the plot

`ylabel('y axis')` places the text 'y axis' to the left of the plot's vertical axis

`text(x,y, 'my text')` writes 'my text' into the current plot at position x,y

Greek letters inside text quotes use: backslash letter, such as `'\alpha'`

Super/subscripts inside text quotes use: _ or ^, such as `'V_1, x^2'` for $V_1$, $x^2$

***Three Dimensional Plots***

`[x,y] = meshgrid(vx,vy)` creates matrices x, y spaced according to the vectors vx, vy

`surf(x,y,z)` creates a surface plot using matrices x,y,z (x,y created using `meshgrid`)

`mesh(x,y,z)` creates a wireframe plot using the same syntax as `surf()`

***Handle Graphics***

`get(gcf), get(gca)` returns all settable option names for current figure and axes

`get(get(gca, 'children'))` returns all settable option names for current data set

`set(gcf, 'keyword',val)` sets the figure option named 'keyword' to val

`set(gca, 'keyword',val)` sets the axis option named 'keyword' to val

`set(get(gca, 'children'),'keyword',val)` sets the data set option named 'keyword' to val

## LAB PROBLEMS

1. A common function in signal processing is the sinc function, $x(t) = \sin(t)/t$. Using exactly 200 points, plot the sinc function for $t = -20$ to 20. Use a thick blue line, ten times thicker than the MATLAB default. ©®

2. Plot the voltage waveform $v(t) = 2 + e^{-\frac{t}{5}}\cos(2t)$ in the style an oscilloscope would show, including using a grid, for $0 \le t \le 10$ seconds. Label all axes and title the plot. ©®

3. Signals and Systems classes teach how to design and analyze filters that remove signals based on their frequency. One such transfer function is:

   $$H(f) = \frac{100}{100 + j2\pi f}$$

   where j is the imaginary value $\sqrt{-1}$, and $f$ is frequency in Hz. $H(f)$ is complex, so can be viewed in complex polar notation as having a magnitude and an angle. Plot the **magnitude** of $H(f)$ as $f$ varies between 1 and 100 Hz. To do this, use `logspace()` to create $f$, and plot the result in a Bode-style plot with the vertical axes in dB and the horizontal axis logarithmically scaled. ©®

4. Controls classes teach how to analyze systems that could control a robot arm's horizontal motion with varying degrees of damping $\lambda$, as described by the equation:

   $$x(t) = \lambda e^{-\lambda t}$$

   Create 3 sets of line plots on the same set of axes showing the effect of $\lambda =$ 0.5, 1, and 2 over the range of $0 \le t \le 5$, with 3 different linestyles (e.g. solid, dotted, dashed), so each will be readable on a black and white printout, and create the corresponding legend for it. ©®

5. High-tension power lines use high voltages to transmit power so that currents, and therefore losses, from resistance can be kept relatively low. Typical high voltage power lines operate at 138kV. Even so, voltages decrease over the power line with distance, as some power is lost to wire resistance and becomes heat energy that radiates into the air. A particular power line has measured voltages [138 136 135 134 133 132] kV at

distances [0 4 8 12 16 20] km from the generating station. Create a scatter plot using large markers of your choice showing these values. Label the horizontal axis "distance from generator (in km)" and the vertical axis "Voltage (in kV)", and title the plot "Power Transmission". ©®

6. Create a plot of a blue square wave y1 and red sine wave y2, both with heights ranging from 1 to +1, and extending from $0 \le t \le 4$ as shown. Hints:

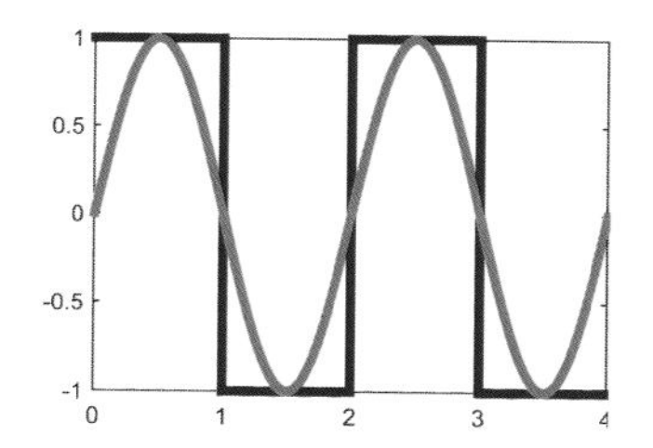


   - Manually specify 8 {t1,y1} pairs to make the square wave, but calculate the sine wave using {t2,y2} with many points to make it smooth.
   - From trigonometry, sin(π t) will fit two cycles in the span of 4 seconds, as shown. ©®

7. Plot the vector of 145 resistor values (from 1Ω to 1MΩ) you saved in Chapter 2 using `subplot()` to create two plot axes over one another. The top one should have the normal (linear) vertical axis, and the bottom one should have a logarithmic vertical axis. Both axes should have the normal (linear) horizontal axis, and both horizontal axes should count from 1 to 145 for each of the standard 5% resistor values. Which plot best shows the data? ©®

8. The complex exponential, $f(\omega) = e^{-j\omega}$, is frequently used in signal processing, but it is difficult to graph directly since for every real value of ω, *f*(ω) evaluates to a complex value. To visualize it, create two subplots, both in the same row, in a single window. Let ω vary from 0 to 8. Hint: you cannot create ω as a variable in MATLAB, so use w as being similar.

   In the first subplot create two different line plots on the same set of axes. One should be the real part of *f*(ω), and the other should be the imaginary part of *f*(ω). Use a legend to tell them apart. Label the horizontal axis with the Greek letter ω and give it the title $e^{-j\omega}$ using superscripts. In the second subplot, plot the real part against the imaginary part. In other words, for this second subplot, use the real part of $e^{-j\omega}$ as the x values and the imaginary part of $e^{-j\omega}$ as the y values, and plot(x,y). Do not label or title the second subplot. ©®

9. Plot the following function:

$$z(x,y)=\frac{2}{e^{(x-0.5)^2+y^2}}-\frac{2}{e^{(x+0.5)^2+y^2}}$$

   Use a grid spanning x and y from -3 to 3 in 0.2 increments. Use the `surf()` command to plot. Hint: To get the x and y to increment in steps of 0.2 instead of their default 1 intervals, you will need to change the arguments into meshgrid. ©®

10. Using the *IEEE Code of Ethics* set out in this chapter, write a one-paragraph answer to the following case study:

    As a Master's engineering student, your thesis involves work on signal processing routines that can identify a type of cardiac (heart) disease based on analysis of electrocardiogram (ECG) signals. ECG signals are the voltage signals measured on the chest surface that propagate from the electrical depolarization of the heart as it beats. Manufacturers X and Y go to court over a dispute of patent rights involving a system that uses a related technology, and you are offered a temporary 2-week job to work as an expert witness for Company X. Your job is to explain to the judge why Y's approach infringes on X's patent claims. As you prepare for the case, you write down the many reasons that this is true; however, you also realize there are several reasons why the opposite argument may be made. At the trial, after you deliver your report, the opposing lawyer asks you during cross-examination if you have any reasons to doubt your reported findings. Does your ethical obligation to your client, Company X, outweigh your ethical obligation to completely disclose your doubts? Would your answer change if the opposing expert witness testified first, and, when faced with the same line of questioning, answered "of course not," giving the judge a biased view?

# 4 MATLAB PROGRAMMING

## OBJECTIVES

After completing this chapter, you will be able to use MATLAB to do the following:

- Describe the differences between MATLAB scripts and functions, and describe the strengths and weaknesses of each
- Explain what input and output arguments are and how they are used
- Write a function that takes and returns any number of arguments
- Use commenting effectively
- Use logical operators to create relational expressions
- Use if-elseif-else conditional structures
- Create strings with embedded numbers

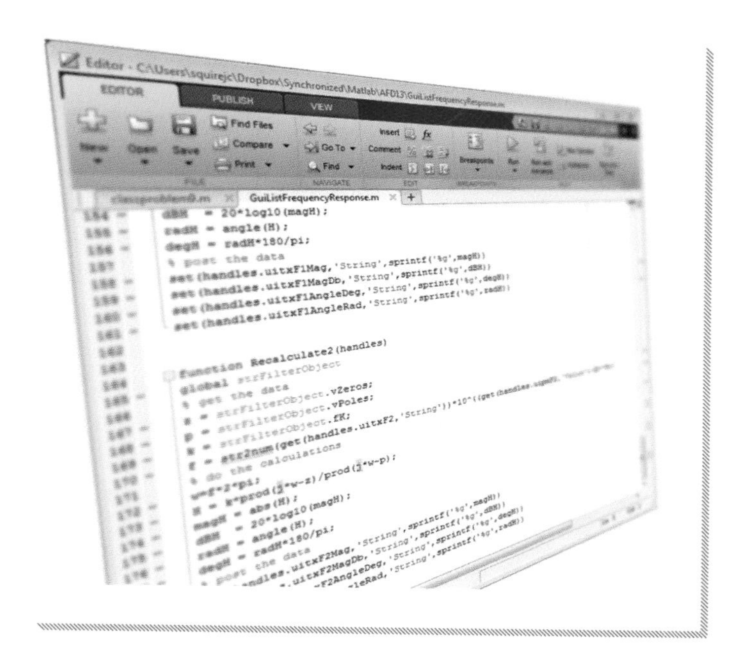

## WHY PROGRAM?

A MATLAB program is simply a collection of commands that one could type from the command line. These commands are grouped together into a file called a program, and is run by typing the file's name into the MATLAB command window. There is nothing one can accomplish from a MATLAB program that cannot be accomplished by typing commands at the command line. So why program? Programs are important because they:

- Simplify repeating a calculation. For instance, instead of having to type the formula `Rparallel = (R1*R2)/(R1+R2)` every time you need it, just create a program called `Rparallel.m`.
- Hide complexity. A user can call a program called `plotFigure2()`, for instance, without having to remember or understand how `plotFigure2()` works.

This chapter introduces programming, which involves two types of text:

- The commands entered at the command prompt >> in the MATLAB command window to run the program, and
- The program itself that is saved in a separate text file.

*Computer Tools* color codes text to differentiate between these two cases: `blue text` represents commands entered in the command window following the >> prompt, and `red text` represents the program code saved in separate file as shown in the diagram below:

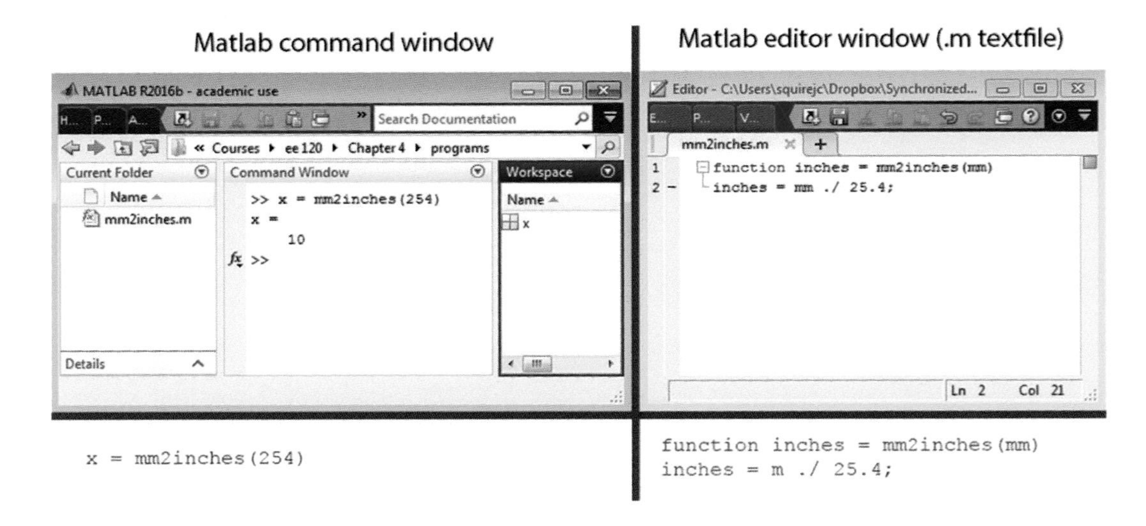

## SCRIPTS

There are two fundamentally different types of MATLAB programs: **scripts** and **functions**. A script file is a collection of MATLAB commands, saved in a file that ends in a .m suffix. When executed, it behaves exactly as if the commands were typed from the command line. All variables created in the MATLAB session before the script is run are available inside the script, and any variables created in the script are left in memory after the script finishes. This means that the user will need to remember the names of the variables that the script expects and ensure they exist before calling the script.

To create a script that computes the value of two resistors in parallel:

1. Decide the names of the function and the variables it acts upon. For example, let the name be `ParallelScript` and have it find the equivalent parallel resistance of variables `R1` and `R2`.
2. Navigate to the directory in which you want to save your program:

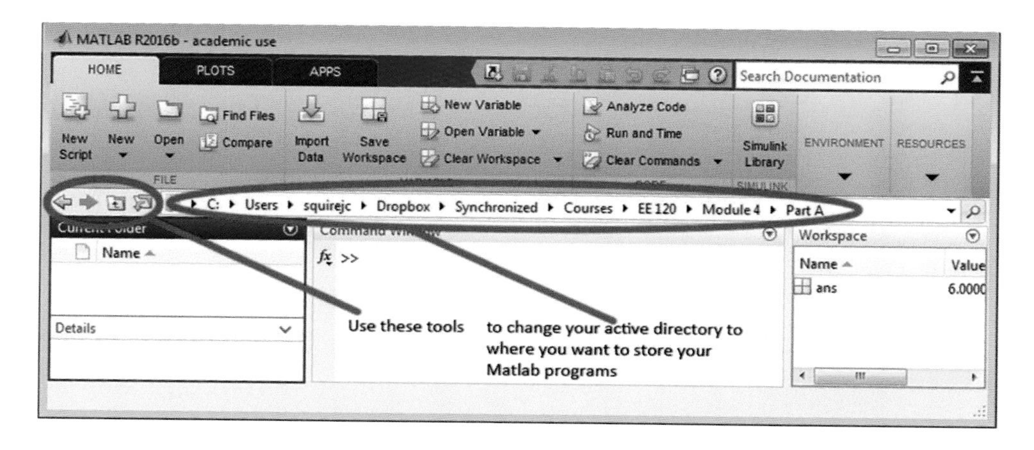

3. In the command window, type "edit" to open a simple text editing window, as shown below:

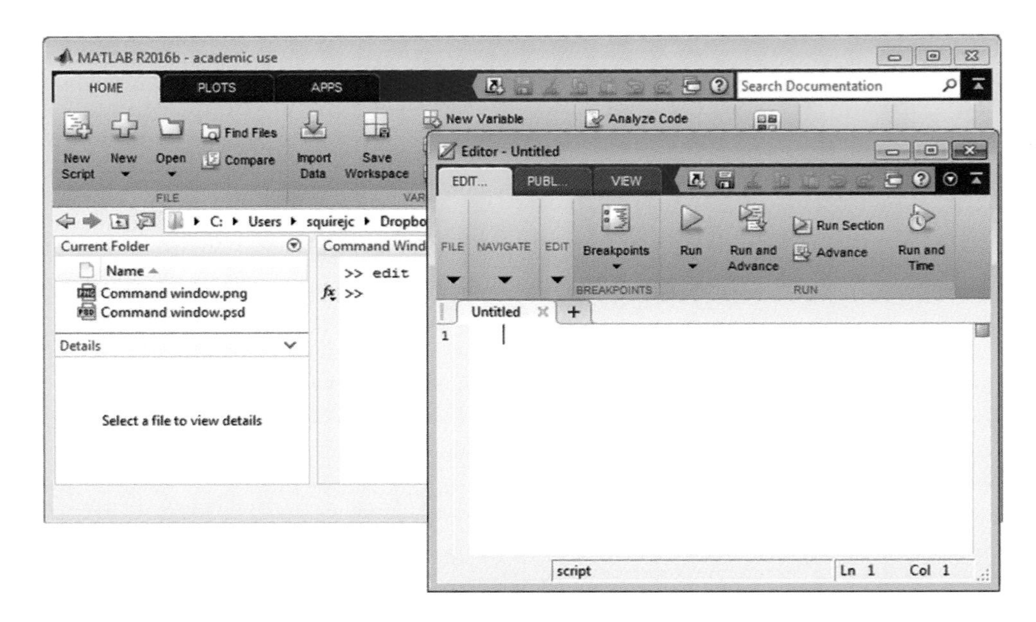


4. Create the script. This is the one-line piece of code:

```
Rp = (R1*R2)/(R1+R2)
```

5. Save the file as `ParallelScript`. MATLAB will automatically append the suffix ".m" to the name to designate it is a MATLAB program. Before you run the script, you must first define the variables it uses in the command window as shown below:

```
R1=10
R2=15
```

6. Now run the script by typing the filename from the command window:

```
>> ParallelScript
```

The script should print:

```
>> Rp = 6
```

When the script is done executing, the variables `R1` and `R2` that were defined before the script ran will remain in the memory workspace along with a new variable `Rp` that was created inside the script.

*NOTE*

Scripts and function names cannot have spaces in them. `ParallelR.m` is a valid MATLAB filename, whereas `Parallel R .m` is not.

*PRACTICE PROBLEMS*

1. Create a script called `CurrentSolver` that expects to have the variables `V` and `R` defined. It should create a variable called `I` and set it equal to the current defined by Ohm's Law: $V = IR$. ©

## FUNCTIONS

Functions, like scripts, are programs consisting of collections of MATLAB commands stored in an .m file. But functions correct two major problems with scripts involving how variables are passed into and out of the program, and how variables created inside the program are handled after the program ends.

***How Scripts Work***

Scripts require that variables be pre-defined with specific names before the script is run, such as R1 and R2 in the above example. Then the script is run, and the script creates variables to store the result, Rp in the above example. A more complicated script may create intermediate variables to help work out the final answer. All of these variables still exist when the script exits – the output variable that holds the answer and any intermediate variables created to help compute the answer.

***Script Problems***

Scripts have two problems. First, the script user must know the name of the input variables, and the names of the new variable that the script creates to hold the answer. If you use someone else's script, you will have to open up the script in an editor to figure out what the input and output variables are called. If the script happened to use the same output variable name that you were already using, your data would be overwritten by the script.

Another problem is that the script may need to create many intermediate variables to help compute the answer. All of these intermediate variables remain in memory after the script exits, potentially cluttering up your workspace with many undesired temporary variables. If any of these variables happen to have the same name as variables you were using, your variables would be overwritten. Both of these problems can be avoided by using functions.

***How Functions Work***

Functions are written the same way that scripts are written, except that they add an extra first line. This extra line declares the program is a function, defines the outputs, and defines the inputs. This is called the **function declaration**, and it will be discussed in detail in the following section.

Functions are called differently than scripts. Scripts are called by first defining the input variables, and then by running the script. Functions do not require pre-defining variables; variables are passed into and out of the function using the following syntax:

```
outputVariables = functionName(inputVariables)
```

You have used this already with built-in functions, like these:

```
y = exp(-6)  % The input variable is -6 and y is set
             % to the function's result
x = sin(pi)  % The input variable is π and x is set
             % to the function's result
```

Here is a comparison of the `ParallelScript` program as a function, now called `ParallelFunction`:

<table>
<tr><th>Script</th><th>Function</th></tr>
<tr><td>Script Program<br><code>Rp = (R1*R2)/(R1+R2)</code></td><td>Function Program<br><code>function Rp = ParallelFunction(R1, R2)</code><br><code>Rp = (R1*R2)/(R1+R2)</code></td></tr>
<tr><td>Using the Script<br><code>R1 = 10</code><br><code>R2 = 15</code><br>The script creates the variable<br><code>Rp = 6</code></td><td>Using the Function<br><code>ParallelFunction(10,15)</code><br>returns 6 to the command window. Since it was not assigned to a variable, it is assigned to the default variable ans:<br><code>>> ans = 6</code></td></tr>
</table>

Notice that the script requires the input variables to be defined R1 and R2, and creates the output variable that is always named Rp. Unlike the script, the function does not require input variables to be defined at all. In this example, the variables were passed directly in, and passed directly out. One could have defined them using the same names as the function declaration used, as follows:

```
R1 = 10
R2 = 15
Rp = ParallelFunction(R1, R2)
```

But the variables also could have been defined as anything because the function is not aware of any variables in the command window, except those passed by the input arguments. For example, the following commands will work just as well as those in the previous example:

```
FirstR = 10
SecondR = 15
Output = ParallelFunction(FirstR, SecondR)
```

In fact, the input variables do not need to be separately defined, as this equivalent example shows:

```
Output = ParallelFunction(10,15);
```

### *How the Function Runs*

When a function is called by the user in the command line, like so:

```
ParallelFunction(10,15)
```

the function begins to execute. It begins with the first line:

```
function Rp = ParallelFunction(R1, R2)
```

This first line tells the function to create temporary variables R1 and R2, and to assign them the values that were passed into the function (i.e. 10 and 15, respectively). Unlike a script, these temporary variables R1 and R2 will not stay in memory once the function exits, and they will not interfere with any other variables in the command window, even if the command window already holds variables of the same name.

Except for the way variable names are handled, the rest of the function works the same as a script, or as if it were typed directly in the command line. For example:

```
Rp = (R1*R2)/(R1+R2)
```

creates a temporary variable Rp and assigns it the result of the computation, with the temporary variables R1 and R2. Here the variable Rp is not left in memory once the program exits, and if the command window had a variable named Rp, the temporary variable would not interfere with it. Inside the function, no variables from the command window are accessible.

After the last line of the function is run, the function goes back to its declaration:

```
function Rp = ParallelFunction(R1, R2)
```

and returns Rp to the command line. This can be viewed if the user calls the function in the command line, as follows:

```
ParallelFunction(10,15)
>> 6
```

Or it can be set to a variable of the user's choice, like so:

```
Rout = ParallelFunction(10,15)
```

This creates the variable Rout and sets it equal to 6. In either case, the temporary variables R1, R2, and Rp defined inside the function will not be left in the command window's memory, and if the command window had already defined variables R1, R2, and Rp, they would not be changed once the function ran.

In summary, functions, unlike scripts, pass whatever variables they need in and out using the function declaration, eliminating the possibility of variable name conflicts. Any variables created while a functions executes will be removed once the function ends, leaving only the variable(s) that the function was designed to return. This behavior of the function is so preferable to that of scripts that we will use functions exclusively from this point onwards in *Computer Tools*.

### *PRACTICE PROBLEMS*

2. Save the following script to `MyScript.m`. It takes three variables: Isource, R1, and R2, and computes current I1.
   ```
   Rp = (R1*R2)/(R1+R2);
   V = Isource*Rp;
   I1 = V/R1;
   ```
   Remove all variables from MATLAB's memory by issuing a `clear` command. Then initialize the variables and run the script, like so:
   ```
   R1 = 10;
   R2 = 15;
   Isource = 5;
   MyScript
   ```
   What are the names and values of the variables in memory after the script is run? These can be viewed from the Workspace panel. ®
3. The above program is re-written as the function below and saved in the file `MyFunction.m`:
   ```
   function I1 = MyFunction(R1, R2, Isource)
   Rp = (R1*R2)/(R1+R2);
   V = Isource*Rp;
   I1 = V/R1;
   ```

Remove all variables from MATLAB's memory by issuing a `clear` command. Then run the function as follows:

```
R1 = 10;
R2 = 15;
Isource = 5;
Ioutput = MyFunction(10,15,5);
```

What are the names and values of all the variables that are in the current MATLAB session (i.e., what are the variables that are visible in the Workspace panel) after the function runs? ®

## FUNCTION ARGUMENTS

The variables passed into or out of functions are called **arguments**. We have seen that these are declared in the first line of the function declaration, e.g.:

```
function outputArguments = functionName(inputArguments)
```

This declaration ensures that the variable outputArguments will be created inside the function, and will set it equal to computed result. Let us examine, then, how any number of input arguments can be passed into the function and any number of output arguments can be passed out of the function.

### *1 Output Argument*

This is the most common case. Examples include the built-in function `exp()` and `sin()`. The function is declared as follows, where input arguments are abbreviated with "…":

```
function outputArgument = functionName(…)
```

For example, a function that converts millimeters to inches could be defined as:

```
function inches = mm2inches(mm)
inches = mm ./ 25.4;
```

Notice that using `./` rather than just `/` means the function `mm2inches()` will work if given a vector *or* a scalar. The above function is saved as mm2inches.m and can be called from the command line using the following syntax:

```
myInchesVariable = mm2inches(254)
```

After calling this function, `myInchesVariable` holds 10.

***2 or More Output Arguments***

Sometimes you will need to return more than one variable. An example of this is a program that converts mm to inches and feet. The declaration for multiple output arguments wraps the output variable names in square parentheses, like so:

```
function [output1, output2] = myFunction(...)
```

For example, a function that converts millimeters to feet and inches could be written as:

```
function [feet, inches] = mm2ftin(mm)
inches = mm ./ 25.4;
feet = floor(inches/12);
inches = inches - feet*12;
```

*RECALL*

The `floor()` command returns the integer part of its argument. `floor(3.7)` for instance returns 3.

The above function must be saved as the file mm2ftin.m and can be called from the command line using the following syntax:

```
[myfeet, myinches] = mm2ftin(5000)
```

Now the variable myfeet equals 16 and myinches equals 4.8504. Note here another power of using functions. Since you previously created a `mm2in()` function, you could change the second line in `mm2ftin()` from:

```
inches = mm ./ 25.4;
```

to

```
inches = mm2in(mm)
```

which simplifies your program by calling a function from within your function.

### *0 Output Arguments*

Occasionally it is useful for a function to not return any arguments, such as when a function just displays a plot. An example you have already encountered is the built-in `plot(x,y)` command. The declaration for a zero output function is:

```
function myFunction(…)
```

For example, the following function creates an artistic-looking plot and returns no arguments to the command line:

```
function coolPlot()
plot(fft(eye(11)));
```

NOTE: PLOTS AS FUNCTIONS

It is often useful to write custom functions to create plots for reports, especially if they are customized from the default plot options. Reports are often revised, and it will save time to be able to quickly regenerate a plot with updated data or to alter its style.

### *1 or More Input Arguments*

Variables are passed into functions in the first-line function declaration statement to the right of the equals sign, as shown below for three input arguments:

```
function outputs = myFunction(input1,input2,input3)
```

When this function is called from the command window using the syntax below:

```
x = myFunction(1,2,3)
```

then variables called input1, input2 and input3 will be created and set equal to 1, 2, and 3 respectively. The function can also be called using the syntax:

```
x = myFunction(a,b,c)
```

Whatever the variable a, b, and c hold will now be assigned to input1, input2, and input3. Variables a, b, and c will not be visible inside the function, but input1, input2, and input3 will be. Once the function ends, input1, input2, and input3 will be destroyed. Any changes the function makes to variables input1, input2, and input3 will not propagate outside the function.

***0 Input Arguments***

Occasionally a function is needed that does not require any inputs. A function that always returns a random capacitor value, for instance, could be defined using the following syntax:

```
function randCap = RandomCapacitor()
```

Note there is nothing inside the parentheses since no variables are being passed into the function. It is called from the command line as follows:

```
x = RandomCapacitor();
```

***Arguments Can Be Scalars, Vectors, or Matrices***

Both input and output arguments can be scalars, vectors, or matrices. For instance, take a custom simultaneous equation solver that takes the A coefficient matrix and the b coefficient vector described in Chapter 2, and computes the result using the following function:

```
function result = Simultaneous(A,b)
result = A\b
```

The inputs, A and b, are a single matrix and vector, respectively, although both could contain thousands of elements.

PRACTICE PROBLEMS

4. Write the function declaration (that is, write the first line of the function *but not the code that calculates the answer*) for a function that computes the maximum frequency of a square-wave generated by a microcontroller given the frequency of the microcontroller's clock. Name the function `clock` and have it take one input called frequency and one output called squarewave. ©
5. Write the function declaration for a function that takes one vector of values, such as a list of all standard 5% resistor values,

and a second input that describes how many random resistor values are requested, and returns a vector of that many random resistors. Only write the function declaration, not the code to solve the problem. Call the function `problem5` ©

## CREATING YOUR OWN FUNCTIONS

There are several steps to creating your own functions:

1. Change the MATLAB working directory into the folder that you wish to hold your functions. This is done in the same way as described under the Scripts heading on page 115.
2. Decide the names of the input and arguments since the function will only have access to these variables. Use this information to write the function declaration, which is the first line of the function.
3. Then decide what commands one would type at the command window to solve the problem using these input variable names. These commands form the remainder of the function. As part of this section, consider if the input arguments will only be scalars, or could be vectors or matrices. Check that the output variable is assigned in the function.
4. Save the function as a file of the same name as declared in the function definition. MATLAB will add a .m suffix.

*Example*

An earlier Practice Problem involved a script to use Ohm's Law $V = IR$ to solve for the current given a voltage and a resistance.

*Solution*

Input arguments: V, R
Output arguments: I

Function declaration: `function I = ISolveFunction(V, R)`
Commands using only the arguments: `I = V./R;`
Note that using "./" instead of "/" allows either or both V and R to be vectors.
Check: output variable I is assigned. Save as ISolveFunction.m:

```
function I = ISolveFunction(V, R)
I = V./R;
```

### PRACTICE PROBLEMS

6. Create a function called `SeriesResistance` that takes two resistor values and returns their equivalent series resistance (which is their sum). ©
7. Create a function called `ParallelResistance` that takes two resistor values and returns their equivalent parallel resistance (which is their product divided by their sum). ©

## COMMENTING PROGRAMS

The percentage sign % is used in MATLAB to denote a comment. All information on the line following a % symbol is ignored. Use this to insert comments to help you understand what your program does or to help break very large programs into blocks. Continuing from the previous example:

```
function I = ISolveFunction(V, R)
I = V./R;     % this is Ohm's Law V=I R solved for I
```

For short, 2-line programs like this you may not see the need for commenting, but as your programs grow to tens or hundreds of lines of code, comments become very important.

## CREATING HELP

You have used help with built-in functions, such as:

```
help linspace
```

You can build help text into your functions as well. This is done by inserting comments on the second line, immediately under the function declaration. All comments between the function declaration and the first line of executable code are returned to the command window when the user types

```
help functionName
```

where functionName is the name of the function in the function declaration, which is the same as the name of the function's file.

Building on the previous example:

```
function I = ISolveFunction(V, R)
% ISolveFunction finds current given voltage and resistance
% I = ISolveFunction(V, R)

I = V./R;    % this is Ohm's Law V=I R solved for I
```

If this is saved as ISolveFunction.m, when the user types:

```
help ISolveFunction
```

The command line window will return:

```
ISolveFunction finds current given voltage and resistance
I = ISolveFunction(V, R)
```

Note the extra blank line that was added inside the program. MATLAB ignores blank lines, so they can help to organize your program into logical blocks, like the help and code areas in this example.

### PRACTICE PROBLEMS

8. Create a commented function `F2C()` that converts a temperature in Fahrenheit to a temperature in Celsius. It should run using the command `F2C(32)`, for example, which will return 0. The conversion formula is $C = \frac{5}{9}(F - 32)$. ©
   The function should implement help; when the user types:
   ```
   help F2C
   ```
   the following is returned to the command line:
   ```
   F2C returns a temperature in Celsius
   given a temperature in Fahrenheit.
   ```

## MORE COMPLEX FUNCTION EXAMPLES

Examine the following functions to develop your programming ability.

***Example: Complex parts in polar form using radians to complex***

Complex numbers in Electrical Engineering are often given in complex polar form, written with a magnitude and an angle that can be in radians or degrees. These were introduced in Chapter 2 in the Complex Numbers section on page 45; an example is 23∠1.23 (radians) or 23∠45° (degrees). MATLAB requires complex numbers to be entered in rectangular form, i.e. using real parts and imaginary parts like 4+j3 or 6*cosd(45) – j*sind(12). Chapter 2 described how a complex number can be changed from polar form of mag∠angle into rectangular form of realpart + j · imagpart using the following formulae:

```
realpart = mag*cos(angle)
imagpart = mag*sin(angle)
```

Once the real and imaginary parts of the number are obtained, they can be assembled into a single complex number **z** using the command:

```
z = realpart + j*imagpart
```

Note that `cos()` and `sin()` in MATLAB expect angles in radians; if you need to convert using angles in degrees use the `cosd()` and `sind()` functions. A function to do the entire conversion follows:

```
function z = polarRadian2Complex(magnitude, radians)
% polarRadian2Complex assembles the magnitude and
% angle parts of a complex number
% into a single complex number.
%
realpart = magnitude *cos(radians)
imagpart = magnitude*sin(radians)
z = realpart + j*imagpart
```

To use it to input $23\angle 0.5$ radians into variable x, save as polarRadian2Complex.m and call it from the command line, like so:

```
x = polarRadian2Complex(23,0.5)
```

MATLAB will respond with:

```
x = 20.1844 + 11.0268i
```

*NOTE*

Inside the function, the output variable is called "z" but it was called from the command line using a completely different variable "x". This is one strength of using functions – the person using the function doesn't have to know what the function calls its variables internally.

***Example: Plot a lowpass filter's response given a cutoff frequency***

Courses in Signals and Systems will develop the following equation that describes a lowpass filter:

$$H(f) = \frac{1}{1 + j\left(\frac{f}{f_o}\right)}$$

In this equation, $f$ is the input frequency in Hz and $f_o$ is the cutoff frequency in Hz. The filter is designed to pass frequencies less than $f_o$ and block frequencies above it. $H(f)$ is a complex function of $f$, since it has a $j$ term in the denominator.

*RECALL*

A Bode-style plot uses semilogx, and plots the vertical (amplitude) axis in dB. To find the value in dB of a complex number **z**, take 20 $\log_{10}|\mathbf{z}|$.

*Problem*

Create a function called `lowpass()` that takes a cutoff frequency $f_o$ and creates a Bode plot of the magnitude of $|H(f)|$ from $1 \leq f \leq 100$.

*Solution*

Since the function takes one number ($f_o$) and returns nothing (it creates a plot), the function definition line and help lines are:

```
function lowpass(fo)
% lowpass(fo) creates a Bode-style plot of the
% response of a lowpass filter with a cutoff of fo
```

Then create a frequency vector, logarithmically spaced between 1 and 100Hz:

```
f = logspace(0, 2, 50); % 1-100Hz = 10^0 - 10^2
```

Next, compute $H(f)$:

```
H = 1./(1+j*(f/fo)); % since using vectors, use ./
```

Now, change that into dB for the Bode plot (see the Recall inset above):

```
dB = 20*log10(abs(H)); % find the magnitude of H in dB
```

Complete the plot and label it:

```
semilogx(f,dB)
title('Lowpass filter')
xlabel('Frequency (Hz)')
ylabel('Amplitude (dB)')
grid('on')
```

Save the program as lowpass.m and call it from the command line for a 10Hz lowpass filter as follows:

```
lowpass(10)
```

The following Bode-style plot is displayed:

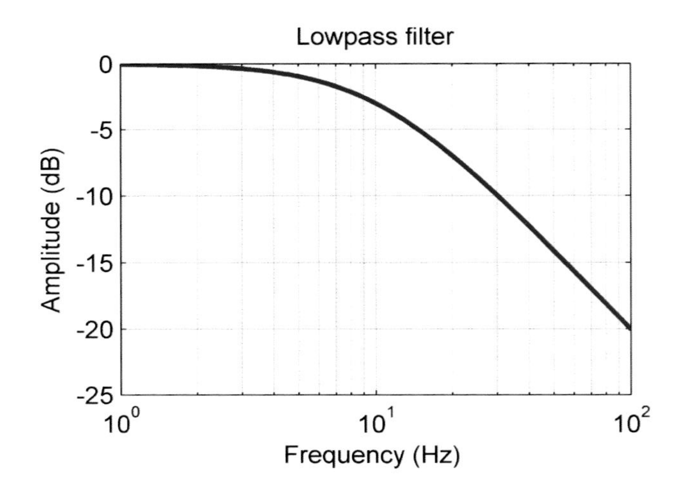


Note that none of the variables created inside the function remain after the function runs.

## PRACTICE PROBLEMS

9. Modify the above function to return the filter response in dB rather than plotting it. For example, using $f_o$=10, a vector of 50 values are returned, the first of which is -0.0432. ©

## SORTING

The MATLAB command `sort()` takes a vector of numbers and returns them in a vector sorted from smallest to largest. For example:

```
vout = sort([4 9 1 -6 0])
```

returns

```
vout = [-6 0 1 4 9]
```

This capability can be useful inside functions that simulate circuit designs over many different random component value variations, and will be used in one of the lab problems.

## PRO TIP: GRADUATE STUDIES IN EE

### Masters

There are two types of masters degree programs: ***industrial*** and ***research***. This is confusing since there are no standardized names for each type of program. Some universities offer both, while others offer only one of the two. Some call one degree an Masters of Science (M.S.) and the other a Masters of Engineering (M.Eng.), while some programs call both an M.S. Examining the graduation requirements will make it clear which course(s) a particular university offers.

- ***Industrial M.S.:*** Industrial M.S. programs prepare students for jobs in industry. They typically require 8-10 classes. A thesis, if required, is small, i.e. the equivalent of 1 or 2 courses. It takes typically one full year to complete, with the associated cost of tuition.
- ***Research M.S.:*** This prepares the student for a research position, and is the first step in earning a doctorate. Coursework varies widely, from as little as 3 courses to as much as required for an industrial masters. A thesis is mandatory and is a major component of the work. It takes longer than an industrial masters, often 18-24 months. This is often not a separate degree program but part of a doctoral degree program, so tuition cost may be waived in return for work as a Teaching Assistant (TA) and Research Assistant (RA), described below.

### Doctorate

Engineering doctorate degree programs are usually sponsored fully through a mix of Teaching Assistantships (TAs) and Research Assistantships (RAs), which typically provide tuition remission and a modest stipend to pay for food and basic accommodations.

In the first year, the student usually works as a TA and teaches undergraduate students. During this year, the doctoral candidate is expected to seek longer-term funding from professors as an RA. This is done by visiting research labs whose work the student finds appealing. If the student and the principal investigator (or PI, the laboratory's lead professor) think they may be a good match (i.e. the PI believes the student will be an asset to the laboratory, and the student enjoys the type of research work, likes the other RAs, and trusts the professor), then the student begins working part time in the laboratory as a trial run. After a semester of this, it should be clear if the lab is a good fit for the student, and the PI will offer an RA to the student. This will allow the student to drop the TA, keep the tuition waiver, and now be paid to do the research that is required by his or her thesis. The PI becomes the student's primary research advisor.

A less-common, but more prestigious type of funding is a fellowship. A doctoral student holding a fellowship receives tuition remission and a stipend comparable to that of an RA, but the source of funding for the fellowship comes from an outside agency, rather than from the PI who funds the RA. This makes students with fellowships very desireable for PIs, who no longer have to pay for their tuition and stipend.

## RELATIONAL EXPRESSIONS

A logical expression is a statement that evaluates to either "true" or "false". Relational operators are a type of logical operator, and compare two values such as 5 > 4 (true) or 3 ≤ -4 (false). MATLAB returns a 1 to indicate true, and 0 to indicate false. MATLAB has several types of relational operators; some of the most common are listed below:

| Operator | Name | Example | Example Result |
|---|---|---|---|
| `>` | greater than | `(5 > 2)` | 1 |
| `<` | less than | `7 < -6` | 0 |
| `>=` | greater than or equal to | `a = -5`<br>`a >= 6` | 0 |
| `<=` | less than or equal to | `a = 7; b = 9`<br>`a*b <= a+b` | 0 |
| `==` | equal to | `5 == 5` | 1 |
| `~=` | not equal to | `5 ~= 5` | 0 |
| `isequal()` | equal to, works with strings, vectors, and matrices | `a = 'hello'`<br>`b = 'Hello'`<br>`isequal(a,b)` | 0 (capitalization matters) |

Note that one of the most common relational statements, the test to see if two scalar numbers are the same, is not = but rather ==.

In the example below, the function called `password()` tests to see if the input is "secret", and outputs a 1, if so and 0, otherwise:

```
function result = password(input)
% password tests to see if the input is
% the string 'secret'
result = isequal(input, 'secret');
```

### PRACTICE PROBLEMS

10. Modify the above function to take a number as an argument and test to see if it is 999. If so it outputs a 1, otherwise it outputs a 0. Call the function `problem10()`. ©

## LOGICAL OPERATORS

Logical operators operate on logical true (1) or false (0) arguments and are particularly important to ECE students, since binary true/false signals form the basis of all digital computers. Common ones are listed below:

<table>
<tr><th>Operator</th><th>Name</th><th>Example</th><th>Example Result</th></tr>
<tr><td><code>&&</code></td><td>AND. True only if both operands are true.</td><td><code>x = 5;</code><br><code>(x>1) && (x<4)</code></td><td><code>0</code> (only one is true)</td></tr>
<tr><td><code>||</code></td><td>OR. True if either operand is true.</td><td><code>x = 5;</code><br><code>(x>1) || (x<4)</code></td><td><code>1</code> (the first test is true)</td></tr>
<tr><td><code>~</code></td><td>NOT. Changes true to false and false to true.</td><td><code>a = 'Hello';</code><br><code>~isequal(a,'T')</code></td><td><code>1</code> (the strings are not equal)</td></tr>
</table>

If you are familiar with digital gates, the associated digital logic symbols are:

<table>
<tr><th>Operator</th><th>Symbol</th></tr>
<tr><td>AND</td><td>&&</td></tr>
<tr><td>OR</td><td>||</td></tr>
<tr><td>NOT</td><td>~</td></tr>
</table>

As an example, suppose you need to check the validity of numbers in a large dataset of electronically-sampled location measurements. These numbers should all be non-negative integers less than 10, or could be -1 if the sensor is returning an error condition. The following function tests whether a particular number meets these criteria:

```
function out = test(in)
isError = (in == -1);  % 0 if non-error, 1 if error
isInteger = (in == floor(in)); % 0 if non-integer, else 1
isInRange = (in >= 0) && (in < 10);
out = isInteger && (isInRange || isError);
```

*PRACTICE PROBLEMS*

11. Create a function called `problem11()` that takes a scalar (single number) and returns a 1 if the input is both not an integer and is negative. Thus `problem11(pi)` returns 0, but `problem11(-pi)` returns 1. ©

## LOGICAL OPERATIONS ON VECTORS AND MATRICES

It is frequently desirable to perform logical operations on large datasets in vectors or matrices. A logical operation on a vector will return a vector of logical results. For example:

```
v = [-2 -1 0 1 2]
b = (v >= 0)
```

results in

```
b = [0 0 1 1 1]
```

There are four common actions done with such a resulting vector (or matrix) of logical values:

- Determine if any of the logical results are true
- Determine if all of the logical results are true
- Determine how many of the results are true
- Replace the values of the vector or matrix for which the given relationship is true with another number

***Determine if any of the logical results are true***

Use the `any()` logical operator. In the above example,

```
any(v)
```

evaluates to true (returns 1) since some of the values in vector v are non-negative.

***Determine if all of the logical results are true***

Use the `all()` logical operator. In the above example,

```
all(v)
```

evaluates to false (returns 0) since some of the values in vector v are negative.

***Determine how many results are true***

Since logical true values are represented by the numeric value of 1, summing the results will return the total number of operations that evaluate to true. Using the above example,

```
sum(v)
```

or more directly

```
sum(v>=0)
```

will return 3.

***Replace the values of the vector for which the given relationship is true with another number***

This is a very common requirement and has its own syntax. Up to this point, vectors have been indexed with a number like v(4) and matrices have been indexed with a pair of numbers like m(3,4). However, they can also be indexed by a logical vector or matrix of the same size as shown in the following example. To replace the negative elements in the following vector with 9's:

```
v = [-2 -1 0 1 2]
v(v<0) = 9
```

Now `v = [9 9 0 1 2]`

PRACTICE PROBLEMS

12. Create a function called `problem12()` that takes a vector of numbers and returns two variables. The first it returns is a variable called "count" that is the count of all the values of the input vector greater than 10. The second is a vector called "result" and is equal to the input, but with all values greater than 10 replaced with 10. In other words, its function definition is:

    ```
    function [count, result] = problem12(x)
    ```

    and as an example of its use,

    ```
    [count, result] = problem12([-5 20 6 11])
    ```

    causes `count = 2` and `result = [-5 10 6 10]`©

## CONDITIONAL BRANCHING: if-end

Thus far, every MATLAB program we have discussed has executed every line of code in it when run. `if-end` statement blocks are an exception; the code between the `if-end` statement only runs if a logical statement is true, e.g.:

```
if (logical expression that evaluates to true or false)
   % Statements here execute if the logical
   % expression is true.
   % All statements until the closing end are
   % conditionally run
end
```

For example, the following statements duplicate the functionality of `abs(x)`, which takes the absolute value of x:

```
if (x < 0) % if x is negative
   x = -x;
end  % continue program
```

Notice that each `if` statement needs to be paired with a matching `end` statement. Also notice that the statements between the `if` and `end` pairs are indented. Since MATLAB ignores spaces, these indentations are optional, but they make the code easier to read.

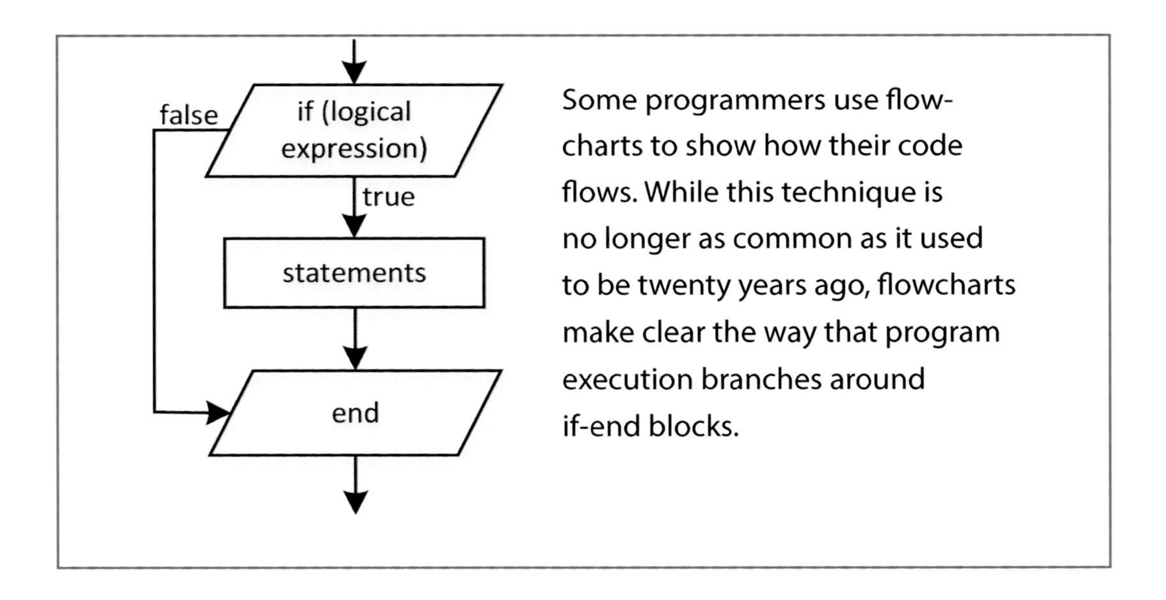


Some programmers use flowcharts to show how their code flows. While this technique is no longer as common as it used to be twenty years ago, flowcharts make clear the way that program execution branches around if-end blocks.

A more complicated example evaluates the `sinc(x)` function, a very common function used in signal processing, which is defined as follows:

$$\operatorname{sinc}(x)=\begin{cases}\dfrac{\sin(x)}{x}, & x\neq 0\\ 1, & x=0\end{cases}$$

This is implemented in the following example:

```
function y = sinc(x)
if x ~= 0
   y = sin(x)/x;
end
if x==0
   y = 1;
end
```

### PRACTICE PROBLEMS

13. Create a function called `problem13` that takes a string and returns a string which is set to the phrase 'unlocked' if the supplied password is 'secret', and 'locked' otherwise. © Hints:
    - Unlike other programming languages, MATLAB uses a single quote ' to denote strings, not double quotes "
    - `==` and `~=` compare numbers; `isequal()` and `~isequal` compare strings, vectors, and arrays.

## CONDITIONAL BRANCHING: if-else-end

In the previous section, a sinc function was developed that ran one set of code if x equaled zero and used a second if statement to run a second set of code if x was not zero. This need, to run one set of code if the logical test is true, and a different set of code if the logical test is false, is so common that it has its own syntax: `if-else-end`.

```
if (logical expression that evaluates to true or false)
   % statements here execute if the logical
   % expression is true
else
   % statements here execute if the logical
   % expression is false
end
```

The `sinc()` function developed in the previous section can be simplified using this syntax, as shown below:

```
function out = sinc(x)
% sinc(x) takes a scalar, returns sinc
if (x==0) % if x is zero
   out = 1;
else
   out = sin(x)/x;
end
```

All conditionally-executing blocks of code can have many statements that run if the logical test is true or false, although this example has just one line of code in each. Remember to end the if-else block with an end statement to mark the code that will run regardless of the if's logical test.

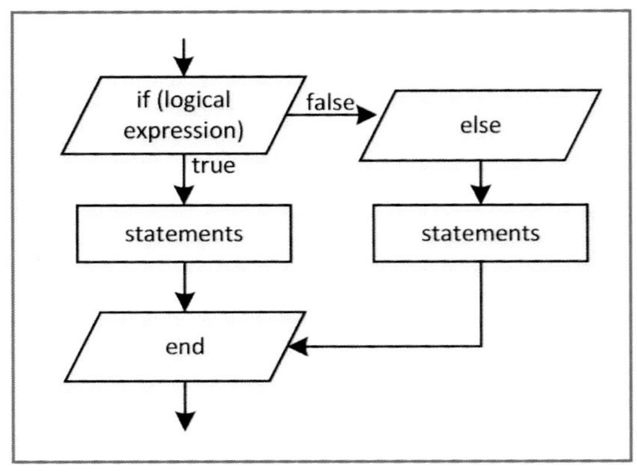


`if-else-end` flowchart showing program execution flow

### PRACTICE PROBLEMS

14. Update the previous `problem13` function to use the `if-else-end` statement, and save it as `problem14`. ©

## CONDITIONAL BRANCHING: if-elseif-else-end

If there is one outcome of a test that requires special code, we use an `if-end` block. If there are two different outcomes that each require special handling, we use an `if-else-end` block. If there are more than two outcomes that require different code blocks to be executed, the `if-elseif-elseif-...-else-end` block works, using as many `elseif` statements are needed for each possible outcome, with a final and optional else statement handling all remaining cases.

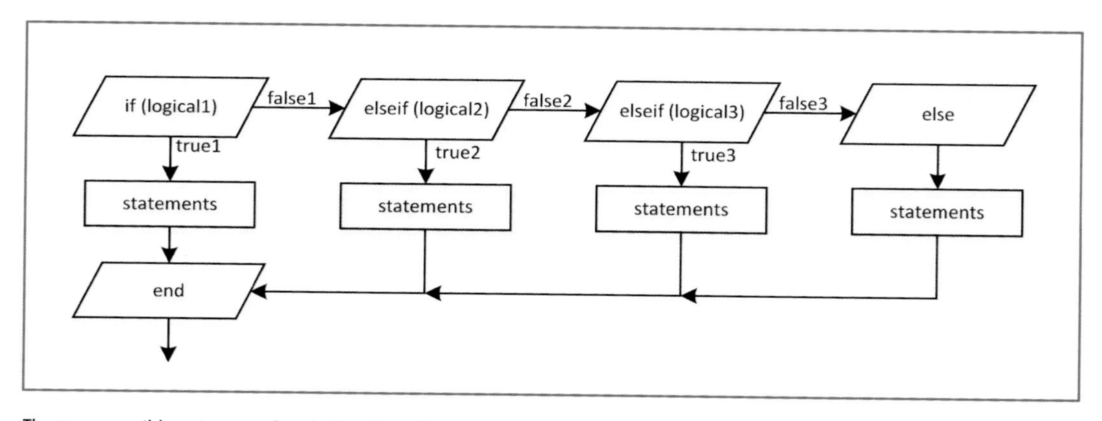


The many possible outcomes of an `if-elseif-else-end` block

As an example, one could generate a random number between 0 and 1, and based on that number provide a number of different responses. For example:

```
function out = isECEfun()
r = rand();
if (r<0.2)
    out = 'always!';
```

```
elseif (r<0.4)
   out = 'sometimes';
elseif (r<0.6)
   out = 'usually, but not on Mondays';
elseif (r<0.8)
   out = 'up until 1:30AM';
else
   out = 'can I change my major to biology?';
end
```

### PRACTICE PROBLEMS

15. Create a function called `problem15` that computes the sign function used in control theory. It takes a number and returns -1 if the input is negative, 0 if the input is zero, and 1 if the input is positive. Use an `if-elseif-else-end` construction. ©

## CREATING STRINGS WITH EMBEDDED NUMBERS

To create a string with both text and embedded numbers in it, use `sprintf()`. For example, the commands:

```
current = 36.5;
s = sprintf('The current is %g Amps.', current);
```

creates a string `s` with the contents `The current is 36.5 Amps.`

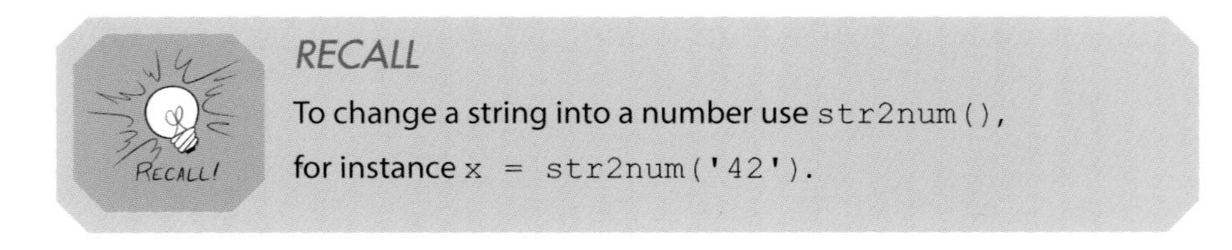


*RECALL*

To change a string into a number use `str2num()`, for instance `x = str2num('42')`.

Each occurrence of the `%g` character inside the single quotation marks is replaced by the number that follows the quotation marks. There can be more than one variable, e.g.:

```
node = 5; voltage = 42;
s = sprintf('Node %g voltage is %g Volts', node, voltage)
```

will cause `s` to hold the string `Node 5 voltage is 42 Volts`

The `sprintf()` command permits extensive formatting. Three of the most common formatting commands are the `\n` command that inserts a new line (i.e. a vertical line space), the `\t` command that inserts a tab to help align tables, and the `%0.2f` command that formats the number with two decimal spaces. As an example, consider the single line below:

```
s = sprintf('Node 1:\t %0.2fV\nNode 2:\t %0.2fV',pi,2*pi);
```

will create a string `s` that displays as the following two lines of formatted text:

```
Node 1:     3.14V
Node 2:     6.28V
```

### *Displaying Strings to the Command Window*

Sometimes you will need to display a result to the command window directly, rather than setting a string variable equal to it. In this case, use `fprintf()`, instead of `sprintf()`, as follows:

```
fprintf('Node 1 = %0.2fV\nNode 2 = %0.2fV\n',pi,2*pi);
prints the following to the command line:
Node 1 = 3.14V
Node 2 = 6.28V
```

Note the trailing `\n` inside the quotation marks; it is necessary because without it, the cursor will be left dangling at the last character written (to the right of the final "V" character in the example just given), and the next character written will continue from that point, rather than being reset to the beginning of the next line as expected.

## PRACTICE PROBLEMS

16. Write the `sprintf()` command (not the entire function) that, given a variable R, prints the statement (for example, if R = 26.2) to the command window ©:

```
The resistance is 26.2 ohms
```

### TECH TIP: WIRE SIZING

Wire sizes in North America are specified by the American Wire Gauge (AWG) standard. Smaller wire diameters have higher wire gauge sizes. General-purpose hookup wire used in electronics prototyping is typically 22 AWG; thicker wire used to carry electricity to kitchen appliances is often 18 AWG, and heavy-duty power extension cables are 14 AWG. Wire is sized as a function of its maximum required current carrying capacity and by the length over which it is required to carry the current. The table below lists common AWG sizes with wire diameter and its recommended current capacity over short distances such as inside appliances:

| AWG | Conductor diameter (in.) | Max current capacity (Amps) |
|---|---|---|
| 12 | 0.0808 | 41 |
| 14 | 0.0641 | 32 |
| 16 | 0.0508 | 22 |
| 18 | 0.0403 | 16 |
| 20 | 0.0320 | 11 |
| 22 | 0.0253 | 7 |
| 24 | 0.0201 | 4 |

Wire is also available in two types: solid and stranded. Solid wire is composed of a single, solid (usually copper) conductor. It is preferred for use in prototyping because it can be pushed into the holes of solderless prototyping breadboards, and for power distribution in buildings because it is easier to connect to receptacles. Stranded wire is composed of several strands of smaller wires bundled together. Stranded wire is more flexible than solid-core wire, and is preferred in applications that may repeatedly bend, such as the power wires connecting to a laptop or an electric iron.

### *PRACTICE PROBLEMS*

17. Write a function called `problem17` that takes an AWG wire size (even sizes 12-24) and returns the following string (using as an example an AWG of 12) "AWG 12 is 0.0808 inches diameter, max current 41A". Use `if-elseif-elseif-end` statements to check the wire size, and use `fprintf()` to print the information to the command window. ©

## TECH TIP: ENGINEERING NOTATION

Engineers use unit postfixes (a letter following a number) to express large or small numbers so that the mantissas are always between 1 and 1000. This is *not* scientific notation: $4.3 \times 10^{-11}$ in scientific notation is written in engineering notation as simply 43 p, for example. These postfixes are in ratios of 1,000. Common ones used in our profession are shown below. Note: Capitalization matters! A 1 MΩ resistor is extremely large; a 1 mΩ resistor is very small. Notice how the numerical part of all the examples are between 1 and 1000:

| **Prefix** | **Name** | **Scaling** | **Engineering example** | **Scientific notation (do not use!)** |
|---|---|---|---|---|
| p | pico | $10^{-12}$ | 12 pF | $12 \times 10^{-12}$ F |
| n | nano | $10^{-9}$ | -434 nV | $-434 \times 10^{-9}$ V |
| μ | micro | $10^{-6}$ | 1.2 μA | $1.2 \times 10^{-6}$ A |
| m | milli | $10^{-3}$ | 150 mH | 0.15 H |
| - | - | 1 | 36 H | 36 H |
| k | kilo | $10^{3}$ | 13 kV | 13,000 V |
| M | Mega | $10^{6}$ | 1 MΩ | $10^{6}$ Ω |
| G | Giga | $10^{9}$ | 96 GB | $96 \times 10^{9}$ B |
| T | Tera | $10^{12}$ | 2 TB | $2 \times 10^{12}$ B |

As you get used to using these symbols you will find they will help you do mathematics spanning large intervals easily. For instance, a 12 μA current through a 4 MΩ resistor generates their product in voltage, which is 12x4 in units of μ (two up from the center) times M (two down from the center), which simply 48 times the center, or 48V. No powers of 10 required!

The `if-elseif-else-end` construct can be used to change numbers into engineering form. Here is an implementation that works for milli, (unity), and kilo prefixes:

```
function s = engform(x)
% engform take a number and rewrites it in
% engineering form
% 0.0032 is returned as 3.2m.
% 12452 is returned as 12.452k
% 98.6 is returned unchanged as 98.6
if (x<1) % values is in milli units
   x = x*1000;
   prefix = 'm';
elseif (x<1e3) % no prefix needed if between 1 and 1000
   prefix = '';
else % value is in a kilo units
   x = x/1000;
   prefix = 'k';
end
s = sprintf('%g %s',x,prefix);
```

Note that `%g` is used in `sprint` for numbers and `%s` is used for strings.

*TECH TIP: RESISTOR COLOR CODES*

The size of a resistor is not related to its resistance but rather to how much power it can dissipate. Many modern ones are in extremely small surface mount packages. The 0201 package, for instance, is about 0.02" x 0.01"; two could fit on the period at the end of this sentence. However, most resistors used for prototyping in solderless breadboards are 1/4W or 1/8W sized and look like this magnified image:

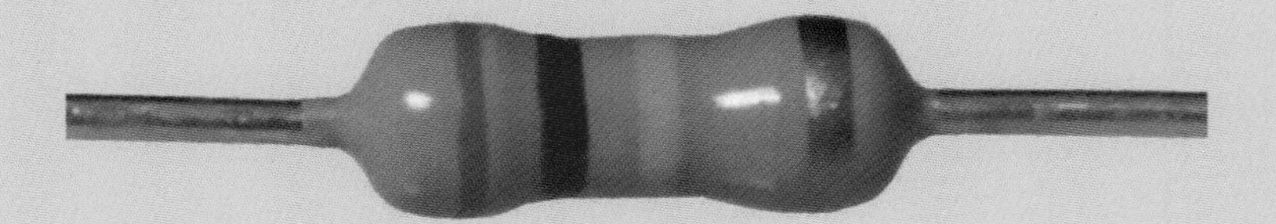

These resistors have four bands of color that with some practice can be read by sight according to the following diagram:

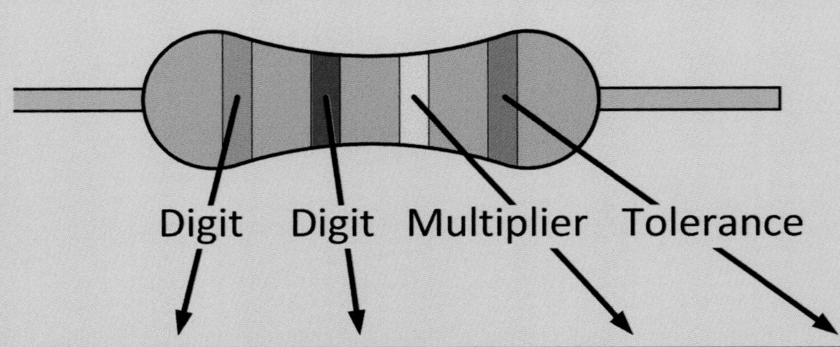


| Color | Digit | Digit | Multiplier | Tolerance |
|---|---|---|---|---|
| Black | 0 | 0 | 1 | |
| Brown | 1 | 1 | 10 | 1% |
| Red | 2 | 2 | 100 | |
| Orange | 3 | 3 | $10^3$ | |
| Yellow | 4 | 4 | $10^4$ | |
| Green | 5 | 5 | $10^5$ | |
| Blue | 6 | 6 | $10^6$ | |
| Violet | 7 | 7 | | 0.1% |
| Grey | 8 | 8 | | |
| White | 9 | 9 | | |
| Gold | | | | 5% |

The above resistor is grey, red, yellow, gold, meaning $82 \times 10^4 = 820\text{k}\Omega$, 5% tolerance.

## PRACTICE PROBLEMS

18. Write a function that takes a resistor digit from 0-9 and returns its color code. Use the `if-elseif-else-end` construct. For example, `s = problem18(4)` sets `s` equal to `'Yellow'`.©

### PRO TIP: INDUSTRY OR GRADUATE SCHOOL?

Even at the start of their studies, EE students often wonder if they should enter industry or graduate school after completing their bachelor's degree. It is a hard decision and perhaps best summarized by listing the benefits of each approach.

***The benefits of going directly into graduate school:***

- Many careers require a graduate degree, such as managing large engineering projects, or performing cutting-edge research.
- Most electrical engineers eventually earn a graduate degree, so it makes sense to do so when:
  - Material learned as an undergraduate is the freshest
  - Additional life responsibilities are less likely to impinge on studies
  - An early start may help boost career advancement significantly.

***The benefits of going into industry first:***

- Graduate school requires a specialization within EE, and it can be difficult to choose without some industry exposure first.
- Since graduate school admissions favor a student's most recent accomplishments, students without high undergraduate GPAs may do well in industry and then gain entry into graduate programs they would not have been accepted into directly out of college.
- New graduates can earn and save significantly during a few years in industry.
- Many employers, especially large ones, pay their employees to get an M.S. in a related field.
- New graduates may find they enjoy the industry position they land with their B.S., thus saving the time, expense, and opportunity cost of pursuing unnecessary graduate studies.

With these many reasons supporting both courses of action it is not surprising that there is no consensus. Statistics indicate that roughly 25% of graduating seniors continue directly to graduate programs in electrical engineering and computer engineering, and about 55% enter engineering-related industry and government positions after graduating. The remaining 20% pursue a variety of other options including military, business, law, and medicine. Of those students who graduate and enter engineering industry, about 75% will continue at some point in their career to graduate school, often after first working three to five years.

Many students seek paid summer internships, often during their rising junior or senior years. Besides earning a substantial amount of money, these often help clarify the choice between industry and graduate school. If the student chooses the former, internships often lead to post-graduation job offers.

## COMMAND REVIEW

### *Utility Functions*

`vout = sort(vin)` returns vector vout of the elements in vin, sorted from smallest to largest

`sum(v)` returns the sum of the values in vector v

### *Relational Statements*

`<, <=` less than, less than or equal to

`>, >=` greater than, greater than or equal to

`==, ~=` equal to, not equal to

`isequal()` isequal works with strings, vectors, matrices

### *Logical Operators*

`a && b` AND, true if both a and b are true

`a || b` OR, true if either a or b is true

`~a` NOT, negates the logical value of a

`all(v)` true if all logical values in vector v are true

`any(v)` true if any logical values in vector v are true

### *Conditional Statements*

`if (a)-end` executes code up to the end if statement a is true

`if (a)-else-end` executes one set of code if a is true, a different set if false

`if (a)-elseif-else-end` multiple tests each with different code sets

### *Embedding Numbers and Strings in Strings*

`sprintf('%g bits',8)` creates a string with "8 bits" in it

`sprintf('%g %s',8, 'bits')` also creates the string "8 bits"

`fprintf('%g bits\n',8)` displays "8 bits" to the command window

## LAB PROBLEMS

1. Create a function called `problem1` that takes two real numbers, a magnitude and an angle in degrees, and returns one complex number whose value in polar coordinates is described by that magnitude and angle. You may need to review the information about complex numbers from Chapter 2. The function definition should be:

   ```
   function z = problem1(mag, degrees)
   ```

   To test it, `problem1(10, 30)` should return `8.6603 + 5i`

2. Create a function that does the inverse of the above problem; it takes a single complex number **z** and returns its magnitude and angle in degrees. The function definition should be:

   ```
   function [mag, degrees] = problem2(z)
   ```

   To test, `[mag,degrees]=problem2(1+j)` should return:

   ```
   mag=1.414, degrees=45
   ```

3. A particular class has an overall semester grade that is entirely dependent on three equally-weighted tests. Write a function that takes the first two test grades, and a desired overall class grade, and returns what the last grade must be in order to earn the given desired overall grade. The function definition should be:

   ```
   function g3 = problem3(g1, g2, gdesired)
   ```

   Testing with `problem3(85, 92, 90)` should return `93`.

4. If you have two resistors you can use them individually or in combination to get four different values of resistance. Write a program that, given 2 resistors R1 and R2, outputs a vector with all possible values of resistance in sorted order (that is, R1, R2, and their value in series, and their value in parallel, sorted from smallest to largest). The function definition is:

   ```
   function Rlist = problem4(R1, R2)
   ```

   Testing it with `Rlist = problem4(10,15)` should return:

   ```
   [6, 10, 15, 25].
   ```

5. More challenging: do the same problem as above, but using three resistors. There are many different ways to combine them (for instance, R2 alone,

R1 in series with R3, R1 in series with (R2 and R3 in parallel). This is a tough problem! See how many unique ways you can combine them – the more unique combinations, the higher your score. Sort the resulting values from lowest to highest. The function definition should be:

```
function Rlist = problem5(R1, R2, R3)
```

Testing it with `Rlist = problem5(30,70,420)` should return:

```
[20 21 28 28.2692 30 60 60.5769 70 80.7692 90 98
 100 420 441 450 490 520]
```

6. Microcontrollers frequently use 8 bit digital values which are integers between 0 and 255, inclusive. Create a function that takes a vector of input values and returns its closest integer between 0 and 255. The function definition should be:

   ```
   function result = problem6(in)
   ```

   Testing it with `problem6([-4.2 0 0.3 2.4 25 254.5 255 999])` should return:

   ```
   [0 0 0 2 25 255 255 255]
   ```

7. Write a function that improves `engform` to work with all of the given prefixes. That is, it should take an argument like 0.0001235 and return '123.5u'. (Use "u" for "μ"). The function definition should be:

   ```
   function s = problem7(x)
   ```

   Hint: build it using the starter code in the Tech Tip: Engineering Notation.

8. Challenging!: Write a function called `problem8` that takes resistance values that is an integer from 1 to 1,000,000, and outputs a string of 4 colors that correspond to the resistor's color code for a 5% tolerance resistor. Do not capitalize the colors, and put a space between them. The function definition should be:

   ```
   function s = problem8(x)
   ```

   Testing it with `problem8(10000)` should return the MATLAB string:

   ```
   'brown black orange gold'
   ```

   Hint: use `if-elseif` statements to print the color strings. There are other techniques that can do this more elegantly not yet covered (e.g., by using switch statements, cell arrays, or string arrays).

# PROGRAMMING II: LOOPING

5

Looping is a powerful programming technique that allows MATLAB users to execute a set of code multiple times, or over every value in a vector. For example, if you need to run a simulation in which a filter's response to variations in resistor tolerance is calculated and displayed for 1,000 random perturbations in resistances, or calculate the output of a circuit whose input is a complicated input waveform sampled at 1,000 different times, the `for-end` loop makes such applications possible.

## OBJECTIVES

After completing this chapter, you will be able to use MATLAB to do the following:

- Program using `for-end` loops
- Understand how to index vectors within `for` loops
- Nest loops inside each other
- Create functions that call other user functions
- Program using `while-end` loops
- Understand how to preallocate memory to speed program execution

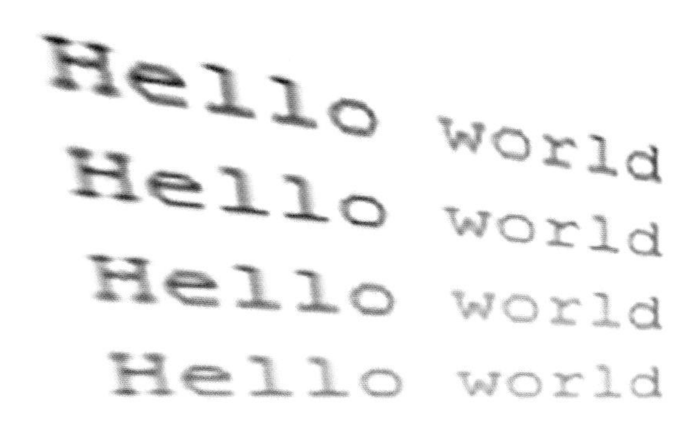

## disp()

Previously, `fprintf()` was introduced as a function to print text and numbers to the screen, possibly using formatting characters including tabs and newlines. A simpler command that can print either pure string text or a variable containing a number or vector, but not combined text with numbers, is `disp()`. For example:

```
disp('Hello world')
v = [1 2 3 4]; disp(v)
```

will cause the following to print to the command window:

```
Hello world
1 2 3 4
```

Use `disp()` instead of `fprintf()` when you only need to print a simple variable. It's frequently helpful when debugging to put in a `disp()` statement to check how the code is evaluating inside a function.

## for-end loop

The `for-end` loop is used to repeat a section of code for a known number of times, as in the example below:

```
for i=1:10
  disp('ECE is fun');
end
```

The `for-end` loop requires the creation of a loop variable, here called `i`, and a loop vector, here the set of numbers from 1 to 10. The loop executes as many times as there are entries in the loop vector, and for each iteration the loop variable takes on a successive value of the loop vector. There can be any number of command lines inside the `for-end` loop. Notice how these commands are typically indented to make them easier to read, much like the commands inside the `if-else-end` block are indented.

For example:

```
for i=1:5
   disp(i)
end
```

will print the following to the command window:

```
1
2
3
4
5
```

The code below makes a list of the squares and cubes of several even numbers:

```
for x=2:2:6
   disp([x  x^2  x^3])
end
```

The above code prints the following to the command line:

```
2     4     8
4    16    64
6    36   216
```

## PRACTICE PROBLEMS

1. What does the following function do? Try to figure it out before you run it to check. Pass in a value between 5 and 10 to check your understanding.

```
function problem1(n)
x=1;
for i=1:n
   x = x+x;
   disp(x)
end
```

2. Fill in the missing part below, labeled ***, to create a function that takes a single resistor value r, returns nothing, and displays the value of r in parallel with resistors of standard values 10, 15, 22, 33, 47, 68, 82 using a `for` loop. Test it with a value of 30. ©®

```
function problem2(r)
for r2 = [10 15 22 33 47 68 82]
   rp = (*** a function of r and r2 ***);
   disp(rp)
end
```

3. Write a function that takes no arguments, returns no arguments, and prints a two-column list that converts degrees to radians. It should print from 0 to 180 degrees in 5 degree increments. ©

## FOR-END LOOPS INDEXING VECTORS

One of the most common uses of a `for` loop is to index a vector. This permits complex operations to be applied to vectors. For example, a different way to accomplish the previous class problem is in the following function:

```
function ForIndexingVectorExample(r)
vAllResistors = [10 15 22 33 47 68 82];
for i = 1:length(vAllResistors)
   rCur = vAllResistors(i);
   rp = (r*rCur)/(r+rCur);
   disp(rp)
end
```

This example begins by defining a vector holding all standard values resistors called vAllResistors. Then it creates a loop variable i which iterates from 1 to the number of resistors in vAllResistors. For each iteration, it pulls one resistor out of the vector and calls it rCur. Then it finds the parallel combination of rCur with whatever resistor value was passed into the function and displays the result.

Up to this point, vector math has been handled with a single operation, for example multiplication of vector v by a constant c is the single operation: `v*c`. This operates on the entire vector at once and is very efficient, but some operations are more easily understood using a `for` loop that steps through every value inside the vector. For instance, consider a piecewise-linear function called the step function and abbreviated u(t), defined and graphed as follows:

$$u(t)=\begin{cases}0, & t<0\\ 1, & t\geq 0\end{cases}$$

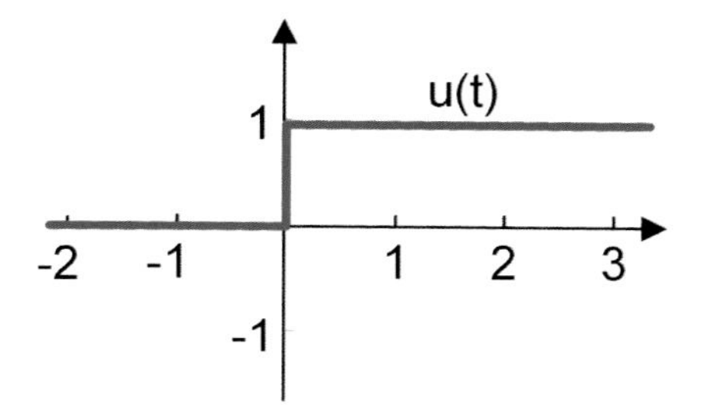


A function that plots this same function is given below:

```
function plotStepFunction() % takes and returns
                             % no arguments
vt = linspace(-2, 3, 1000); % horizontal axis
                             % has 1000 points
vu = zeros(1,1000);     % create vertical data vector
                        % of same size
for i = 1:length(vu)
  t = vt(i);            % t loops through each point
                        % in vector vt
```

```
    if (t<0)            % if t is negative
      vu(i) = 0;        % then vu at that time is 0
    else                % otherwise
      vu(i) = 1;        % vu at that time is 1
    end
  end
  plot(vt,vu)
```

Notice how the `for-end` loop sets t to every possible value of the vt vector, and then individually performs some operation on it – here, checking to see if it is positive or negative, and then creates the output vector accordingly.

The more complex example plot below plots the following mathematical function from $0 \le t \le 30$:

$$f(t)=\begin{cases} \sqrt{10t}, & 0\le t\le 10 \\ 10e^{10-t}, & 10\le t\le 20 \\ t-20, & t\ge 20 \end{cases}$$

```
function plotPiecewiseFunction()
vt = linspace(0,30,1000);    % let t range from 0 to 30
vf = zeros(1,1000);          % fill with 0's
for i=1:1000
  t = vt(i);                 % iterate so t will be
                             % every value in vt
  if (t >= 0) && (t < 10)
    vf(i) = sqrt(10*t);
  elseif (t >= 10) && (t < 20)
    vf(i) = 10*exp(10-t);
  elseif (t >= 20)
```

```
        vf(i) = t-20;
    end
end
plot(vt,vf)
```

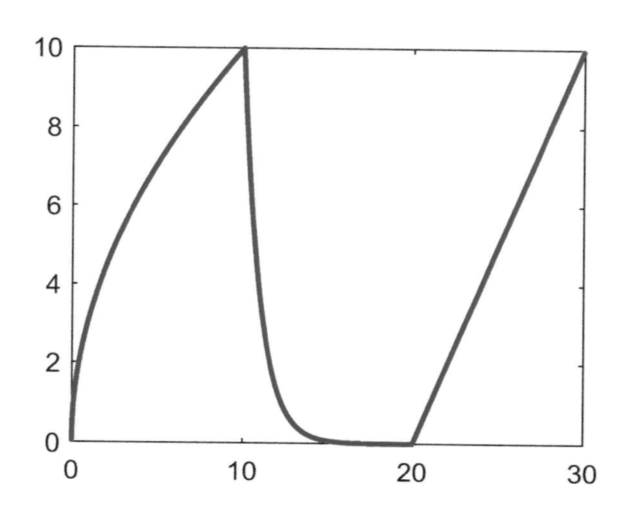


*PRACTICE PROBLEMS*

4. Create a function that plots the following for -5 ≤ t ≤ 5. Plot using 100 points. ®

$$v(t)=\begin{cases}1-t, & t<0\\ e^{t/4}, & t\geq 0\end{cases}$$

### TECH TIP: MONTE CARLO SIMULATIONS

Monte Carlo simulations are commonly used in EE to test how engineering designs respond to random changes in component tolerance. For example, consider the lowpass circuit shown below, which passes low frequencies but tends to block frequencies above $f = \frac{1}{2\pi RC}$:

R
$V_{in}$ C $V_{out}$

Choose R = 1kΩ and C = 1µF to make the filter pass frequencies below 160Hz. How will this cutoff frequency vary given a resistor tolerance of ±5% and capacitor tolerance of ±10%? A graduate-level course in stochastic theory will derive the result exactly, but it is simpler to simulate the results several thousand times, each time with different random resistor and capacitor values.

In MATLAB, `rand` returns a random number between [0,1]. To change this to [-0.05 +0.05] to simulate a 5% resistor tolerance, note there is a desired span of 0.1 in [-0.05 0.05], so `(rand*0.1)` gives the correct span of [0 0.1]. Subtract 0.05 to create the desired range [-0.05 0.05]. Therefore `rand*0.1-0.05` gives a random number between -0.05 and 0.05 for the resistors. Similarly, `rand*0.2-0.1` gives a random number from -0.1 to 0.1 for the capacitors.

```
function result = MonteCarlo(N)
% N is the number of simulations desired
result = zeros(1,N);      % fill result with
                          % N zeros
for i=1:N       % do the simulation N times
```

```
    R = 1000 + 1000*(rand*0.1-0.05);
      % +/-5% resistor tolerance
    C = 1e-6 + 1e-6*(rand*0.2-0.1);
      % +/-10% capacitor tolerance
    f = 1/(2*pi*R*C);
    result(i) = f;
end
```

Run the function from the command window and have it do 1,000 simulations. The results show that most, but not all, of the critical frequencies span from 140Hz to 180Hz, something difficult to derive using methods other than simulation.

```
result = MonteCarlo(1000);
plot(result,'.')
xlabel('simulation number')
ylabel('critical frequency')
```

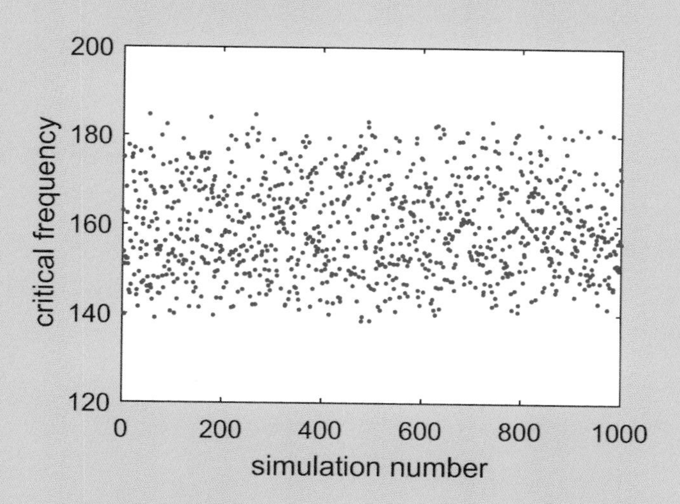


A new MATLAB command that can help visualize the results of Monte Carlo simulations is `histogram()`. This command bins the data and plots a histogram of the results.

```
histogram(result); % plot the histogram
xlabel('critical frequency')
ylabel('number')
```

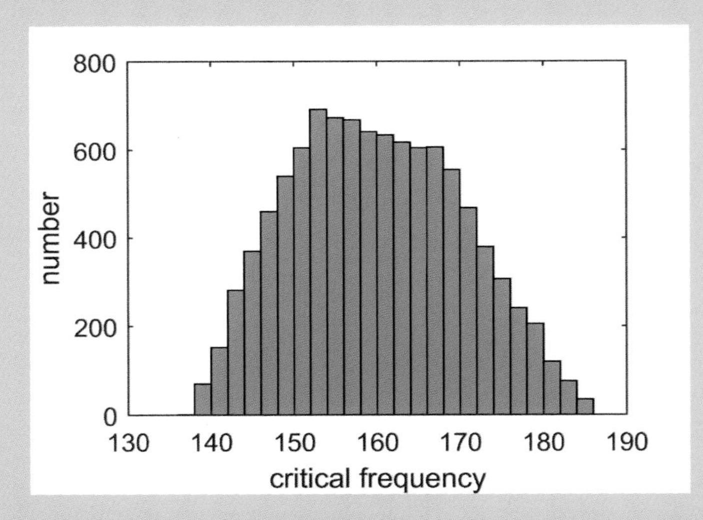


Now it is apparent that although the average value of the filter's cutoff is about 160Hz, which is what would be obtained with resistors and capacitors whose values are precisely as marked, using real-world resistors and capacitors with about a 5% and 10% tolerance, respectively, will result in a spread of cutoff frequencies. One could expect to find cutoffs as low as 140Hz and as high as 180Hz when using components with real-world, imperfect values. Notice how the `histogram` command displays the same information as the previous MonteCarlo function does, but in a more easily-interpreted format.

### *PRACTICE PROBLEMS*

5. Solve the Monte Carlo filter problem on the previous page, but now assume that the capacitors are higher quality 5% versions, and the resistors have a 1% tolerance. Plot the results of 1,000 simulations as a histogram. ®

## NESTED LOOPS

For-end loops are often nested inside each other. For example, if a parts cabinet were stocked with exactly 4 types of resistors, 10Ω, 22Ω, 47Ω, 51Ω, find all possible combinations of these types:

```
function FindResistors()
% FindResistors finds all combos of 2 resistors
% given 4 types
vr = [10 22 47 51];    % 4 types of stocked resistors
for i1 = 1:4        % the first resistor index
  R1 = vr(i1);      % R1 is the first resistor value
  for i2 = 1:4      % choose the second resistor index
    R2 = vr(i2); % R2 is the second resistor value
    fprintf('R1 = %g, R2 = %g\n', R1, R2)
  end
end
```

This code returns the following output:

```
R1 = 10, R2 = 10
R1 = 10, R2 = 22
R1 = 10, R2 = 47
R1 = 10, R2 = 51
R1 = 22, R2 = 10
```

```
R1 = 22, R2 = 22
R1 = 22, R2 = 47
R1 = 22, R2 = 51
R1 = 47, R2 = 10
R1 = 47, R2 = 22
R1 = 47, R2 = 47
R1 = 47, R2 = 51
R1 = 51, R2 = 10
R1 = 51, R2 = 22
R1 = 51, R2 = 47
R1 = 51, R2 = 51
```

### *PRACTICE PROBLEMS*

6. Modify the above code to print the equivalent resistance of all possible values of R1 and R2 in parallel. ©®
7. Is it possible to sort the above result with no code changes other than by simply inserting the sort() command in the correct place? If so, do it and report the result. If not, explain why it is not possible. This problem should clarify when it is helpful to print results to the screen and when it is helpful to return results as a function's output argument.

## USING NESTED LOOPS TO SEARCH FOR **EXACT** SOLUTIONS

Consider a program that searches for integer solutions for the sides of a right triangle, known as Pythagorean triples, satisfying $c = \sqrt{a^2 + b^2}$. The program searches for integer hypotenuses as side *a* ranges from 1 to some given upper bound *N* and as side *b* ranges from 1 to *N*.

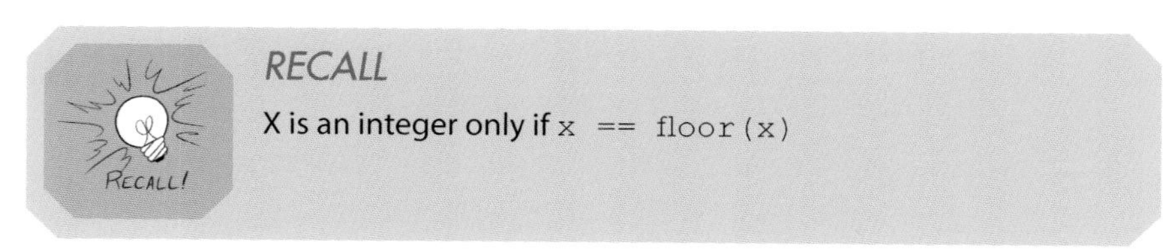


Every time an integer hypotenuse *c* is found, it is printed to the command window, as shown below:

```
function pythagorean(N)
% pythagorean(N) searches for integer sides of right
% triangles for lengths of sides varying from 1 to N
for a = 1:N                     % side a
  for b = 1:N                   % side b
    c = sqrt(a*a+b*b);          % side c, the hypotenuse
    if (c==floor(c))  % a solution is found
      disp(c)
    end
  end  % loop through side b values
end    % loop through side a values
```

### PRACTICE PROBLEMS

8. The above code for `pythagorean()` works, but it only prints the hypotenuse values. Modify the code to print all the sides *a*, *b*, and *c*, and list the results for sides *a* and *b* between 1 and 20. ®

*PRO TIP: NON-ENGINEERING CAREERS*

Many electrical engineering graduates choose to pursue careers other than Electrical Engineering, and find that the quantitative skills they learned as students transfer well. Popular non-engineering careers for EE graduates include:

- **Sales:** There are many opportunities for qualified engineers who are more interested in working with people than design. Although these jobs are really about building trusting relationships, many require engineering-level knowledge about the system being sold, such as when selling substation transformers to provide power to growing communities.
- **Project management:** This is another people-centric position, and it requires the ability to manage teams of engineers during the development phase of a project. Such positions require far more communication and planning skills than design skills.
- **Military:** Increasing automation, especially in the technology-centric branches of the Air Force and Navy, calls for greater numbers of engineers. There are often special ROTC scholarships reserved specifically for engineering majors to help fill this gap.
- **Medicine:** Engineers have higher Medical College Admission Test (MCAT) scores on average than both biology majors and "pre-med" health science majors[1], and the analytical training they receive gives them advantages in certain specialties, including cardiology, neurology, and radiology.

- **Law:** Math-intensive undergraduate curricula are correlated with Law School Admission Test (LSAT) performance; according to 2016 data, engineering students are the fifth-highest performing, on average, with physics majors ranking first[2]. Patent law recruits heavily from engineering majors; an undergraduate degree in engineering alone is sufficient qualification to take the U.S. Patent Bar Exam.

Opportunities abound both within and outside the engineering profession. Which will you choose?

[1] "MCAT and GPAs for Applicants and Matriculants to U.S. Medical Schools by Primary Undergraduate Major, 2016-2017," Association of American Medical Colleges, 2016. www.aamc.org/download/321496/data/factstablea17.pdf

[2] "LSAT Scores of Economics Majors: The 2003-2004 Class Update," *Journal of Economic Education* (Spring 2006): 244-247.

## USING NESTED LOOPS TO SEARCH FOR **BEST** SOLUTIONS

Finding exact solutions, as was done in the previous section, is done by testing with the `==` operator. Sometimes exact solutions do not exist, in which case one must settle for finding the *best* solution. Finding this best solution requires keeping track of the current best solution, and checking through every iteration to see if the current computation is better. Error is often computed using code of the form `error = abs(currentComputation - desired)` since it is the absolute error that needs to be minimized.

For example, if you need to find the value of two standard value resistors from 1Ω to 1MΩ in series, whose sum is as close as possible to a given desired value, enter the following:

```
function [R1best, R2best] = findclosestseries(rDesired)
% create vR with standard value resistors from 1 to 9.1
vR = [1 1.1 1.2 1.3 1.5 1.6 1.8 2 2.2 2.4 2.7 3 ...
   3.3 3.6 3.9 4.3 4.7 5.1 5.6 6.2 6.8 7.5 8.2 9.1];
   % ... wraps the line

% duplicate vR by multiples of 10 to create all
% standard value Rs
vR = [0 vR vR*10 vR*100 vR*1e3 vR*1e4 vR*1e5 1e6];
R1best = 0; % we haven't yet found the best value of R1
R2best = 0; % we haven't yet found the best value of R2
bestError = 1e12; % our initial error is huge

% loop through to find every possible value of R1
for i1 = 1:length(vR)
   R1 = vR(i1);
   % loop through to find every possible value of R2
   for i2 = 1:length(vR)
      R2 = vR(i2);
      Rseries = R1+R2;
      currentError = abs(Rseries - rDesired);
      if (currentError < bestError)
         % we found a better solution
         R1best = R1;
         R2best = R2;
         bestError = currentError;
      end
   end
end
```

### PRACTICE PROBLEMS

9. Change the above program to print out the closest value of two standard-value resistors in parallel to a given desired resistor. Test to find the nearest parallel combination of standard resistor values to match R=40.5Ω. How close is it? ®

## tic, toc

The commands `tic, toc` will time how long it takes MATLAB to execute the commands that separate them. These commands must be either within a function, or cut and pasted into the command window; otherwise, most of the delay between the "tic" and the "toc" will be from your typing speed, not MATLAB's calculation speed limitations.

```
function calculationspeed
% find how long it takes to create two random 50x50
% matrices and multiply them together
tic
a = rand(50,50); % create two random 50x50 matrices
b = rand(50,50);
c = a*b; % multiply them together
toc
```

MATLAB then returns the following on my 2017-era computer:

```
Elapsed time is 0.278700 seconds.
```

The toc command will also return the elapsed time in seconds to a variable.

### PRACTICE PROBLEMS

10. How quickly can MATLAB solve a set of 100 equations with 100 unknowns? This problem was described in Chapter 2, but to avoid typing 10,000 test coefficients, create a random problem. First create a 100x100 matrix of random numbers and save in variable A using the command:
    ```
    A = rand(100,100);
    ```
    Then create a 100 x 1 column vector of random numbers called b. Finally, time how long it takes to solve them using the command:
    ```
    V = A\b;
    ```
    ®

## FUNCTIONS CALLING FUNCTIONS

Functions frequently call other functions inside them, as shown below:

```
function z = polar2complex(mag, angle)
% polar2degrees returns a complex number given
% the magnitude and angle in degrees
z = mag*cosd(angle) + j*mag*sind(angle);
```

The `sind()` and `cosd()` are built-in functions called inside `polar2complex`. User-created functions can be called the same way. Consider the following function that returns the equivalent resistance of two resistors in parallel:

```
function p = parallel(r1, r2)
% parallel returns the parallel resistance
% of two given resistors
p = (r1*r2)/(r1+r2);
```

The parallel function can be called by other user-defined functions. For example the Monte-Carlo routines described on page 164 could call `parallel` repeatedly in its simulations, allowing the Monte-Carlo code to become more readable.

## break

It is sometimes necessary to break out of the middle of a loop before it is completed. For example, this might happen if the loop is searching for some criteria to be met, such as trying to minimize a calculated error below some threshold. Once this is accomplished, the loop has served its purpose; in such a case, you can exit the loop immediately using the break command.

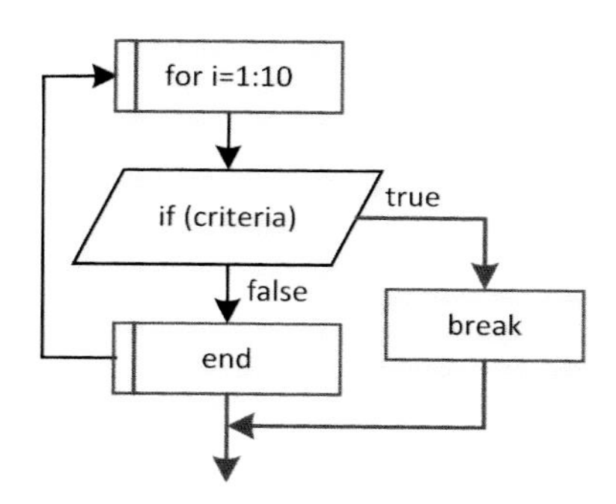


Another example involves searching to see if a given number is prime. One method, given a number n, is to loop over all possible integers between 2 and n to see if any are factors of the given number. The moment any factors are found, the loop exits and reports that the number is not prime. If the loop completes with no factors found, the function reports that the number must be prime. But rather than searching for possible factors from 2 through n, one only needs to search up to the square root of n, since any factor greater than this must be multiplied by a factor less than this.

Define the following helper function and save as isInteger.m:

```
function b = isInteger(x)
b = x == floor(x); % b is 1 if x is an integer,
                   % otherwise 0
```

Last, define the main function to locate primes, and save as isPrime.m:

```
function b = isPrime(n)
b = 1;  % assume it is prime, then check
for f=2:floor(sqrt(n))  % f cycles through
                        % all possible factors
```

```
      if isInteger(n/f) % is f a factor of n?
           b = 0; % change b to 0
           break  % break out of the loop
      end
   end
```

This program would run correctly without the break statement, but it would be slower; if a factor were found, it would continue to run the loop until the end, rather than returning the result immediately.

*PRACTICE PROBLEMS*

11. Change the above code in isPrime to check to make sure that the program has not been running for over 1 second. Do this by starting a timing operation at the beginning of the code using `tic`, and check the elapsed time inside the loop with `toc`. Use the `break` statement to exit if the code has been executing for over 1 second, and if so exit and return -1. This is a useful technique to exit automatically when a simulation does not converge on an answer. ©

## MULTIPLE FUNCTIONS IN ONE M-FILE

In the previous example, a separate helper function called `isInteger` was saved as a separate file, even though its purpose was only to be used within the `isPrime` function. It is possible to pack multiple helper functions inside the same m-file that houses the main function, as long as the helper functions are defined after the main function. These helper functions can only be called by the main function.

For example, re-writing the last example using this method yields a single file to be saved as `isPrime2.m`:

```
function b = isPrime2(n)
b = 1;  % assume it is prime, then check
for f=2:floor(sqrt(n))  % f cycles through
                        % all possible factors
  if isInteger(n/f)     % is f a factor of n?
      b = 0; % change b to 0 to show n is not prime
      break; % break out of the loop
  end
end
%
% helper functions
function b = isInteger(x)
 b = (x == floor(x));    % b is 1 if x is an integer,
                         % otherwise 0
```

Defined this way, the helper function `isInteger` cannot be seen by the MATLAB command line or any function other than `isPrime2.m`.

## while-end loop

The `for-end` command loops for a number of times that is known in advance, unless a `break` statement ends it early. The `while-end` loop will loop until a logical condition is true.

```
while(logical test statement)
  program code to be run
end
```

It is less common than the `for-end` loop command, and like the `for-end` loop command, can be exited early using the `break` statement. It can be used, for instance, in simulations that iterate until an error is acceptably small.

```
function sum=WhileExample(x)
% WhileExample will loop until new additions are < x
```

```
term = 1;
sum = 0;
while (term >= x) % run until the terms are less than x
   term = term/2;
   sum = sum+term;
end
```

This function snippet adds the geometric series $\frac{1}{2}+\frac{1}{4}+\frac{1}{8}+\ldots$ until the last added term becomes smaller than some given number x.

*PRACTICE PROBLEMS*

12. Change the above code so that it sums the sequence $\frac{1}{3}+\frac{1}{9}+\frac{1}{27}+\ldots$ Report the result for summing until the last term added is smaller than 0.00001. ®

## GROWING VS. PRE-ALLOCATING VECTORS

Often a function will return a vector whose elements are calculated inside a loop. In this case, the vector may be either initialized to the full size with a `zeros()` command and then filled inside the loop, or else grown to size inside the loop with its length increasing with every loop. The former method is much more efficient, and is accomplished by reserving the required memory for the vector before the loops begins with a command like this:

```
result = zeros(1,1000);
```

The latter less-efficient method is to grow the size of vector in the middle of the loop, like this:

```
result = [newvalue result];
```

If the resulting vector size is known before the loop begins executing, as is often the case, use the former method. Only if the resulting vector length is not known in advance should the latter method be used.

***Good practice: preallocate memory***

```
function v = GoodProgram(n)
% return v of length n and preallocate v
v = zeros(1,n);
for i=1:n
   v(n) = rand + sin(2*i);
end
```

***Bad practice: grow memory***

```
function v = BadProgram(n)
% return v of length n but grow v to size
v=[];   % define v as being empty
for i=1:n
   newval = rand + sin(2*i); % newval is calculated
   v = [newval v];  % insert newval at start of v
end
```

*DIGGING DEEPER*

The reason why vectors should be initialized with some value like zeros, rather than grown as needed, is because of the way MATLAB grows vectors. If a vector needs to be grown to accommodate a new value, for instance to go from length 100 to length 101, MATLAB must first internally create an entirely new 101 length vector, then copy the original 100 values over to it, then save the additional single value, and then destroy the original length 100 vector. Compare this with the faster alternative of preallocating memory for the final array once, and then replacing the initialized values with the calculated values.

*TECH TIP: BINARY MATH WITH MATLAB*

There are two types of mathematics electrical engineers do with 0 and 1: **logical** and **binary**.

***Logical, or Boolean Math***

These use numbers to represent true (1) and false (0). The logical operations of &&, ||, and ~ model the effects of their counterpart logic gates AND, OR, and NOT gates.

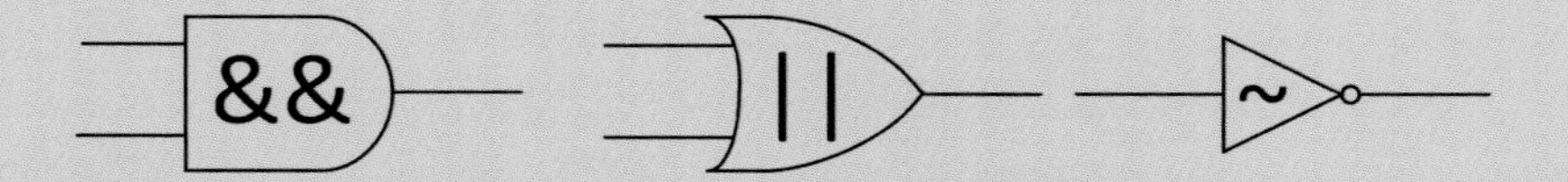


As an example, find the output of the following digital circuit with MATLAB:

The output can be found by substituting the MATLAB logical operations, as follows:

The MATLAB code to calculate the answer given the inputs therefore is ~((1||0)&&1), which evaluates to 0.

***Binary math:*** `dec2bin()` ***and*** `bin2dec()`

Binary mathematics involves calculations with numbers in base 2, but MATLAB shows all numbers in base 10. To convert to base 2 use `dec2bin()`

For example, to convert 77 to base 2:

`dec2bin(77)` returns `1001101`

To convert back into base 10, use bin2dec and put string markers around the binary value

`bin2dec('1001101')` returns `77`

As an example, perform the following operation in MATLAB: $10011_2 \times 1001_2 + 1010_2$:

```
bin2dec('10011')*bin2dec('1001')+
   bin2dec('1010')
```

This returns 181 (in base 10). To see the result in binary, use:

```
dec2bin(181)
```

This returns:

`10110101` (in base 2)

## *PRACTICE PROBLEMS*

13. Use MATLAB to calculate the following logic block ©®

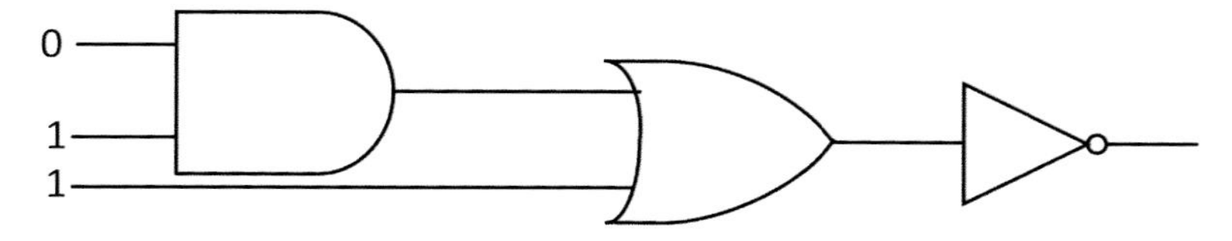


14. Use MATLAB to evaluate the following binary math problem, and report the answer in binary: $(1001_2 + 1110_2) \times 10011_2$ ®

*PRO TIP: FUNDAMENTALS OF ENGINEERING*

An earlier Pro Tip discussed the advantages to taking the optional step of earning the Professional Engineering license. The first step in becoming a PE is to pass the Fundamentals of Engineering (FE) exam. This test is designed to be taken during the senior year, or by recent graduates of an engineering program. There are versions of the FE exam for Electrical, Mechanical, Civil, and other engineering majors; the FE exam includes topics common to all of these majors, such as mathematics, probability, ethics, and engineering economics, as well as topics unique to the selected engineering discipline. The EE exam, for instance, also includes circuit analysis, signal processing, electromagnetics, digital systems, and programming. The exam is composed of two three-hour, computer-graded segments, which are taken with a break for lunch. Certain models of scientific non-programmable calculators are authorized, and a book containing reference equations is provided. You can find out more about the exam, how to purchase a copy of the reference equation book, and how to register for it by visiting NCEES.org, the administering testing agency.

Although scores are reported, it is essentially a pass/fail exam since its only purpose is to serve as one of several steps required to become eligible to take the PE exam. In past years roughly 70% of electrical and computer engineering test-takers earned passing scores, and one can retake the exam as many times as required, up to once in every two-month testing window, and up to three times per year.

Very few undergraduate majors have a national licensing examination as engineers do. Although most electrical engineering careers do not require the PE, it is difficult to determine as a student where your particular career path will lead. Why not take advantage that national licensure exists for electrical engineers, and gain maximum career flexibility and a distinguishing résumé line by planning to take the FE as a senior?

## COMMAND REVIEW

***Looping Functions***

`for-end` for i=1:10, fprintf('Loop %g\n', i); end

`break` breaks out of a loop

`while(condition)-end` loop until the logical condition is met, then exit

***Utility Functions***

`tic, toc` prints elapsed time

`tic, t=toc;` saves elapsed time in seconds in variable t

`disp()`

***Binary Functions***

`dec2bin(73)` converts a base 10 value into base 2

`bin2dec('101')` converts a base 2 number (in quotes) into base 10

## LAB PROBLEMS

1. Write a MATLAB program that takes vector t, and returns a vector v, according to the formula:

$$v(t)=\begin{cases} 1+t, & t\le 0 \\ 2e^{-t}-1, & t>0 \end{cases}$$

   Use a `for-end` loop in your solution. The function definition line is:

```
function v = problem1(t)
```

   Test your code by verifying in the command line that `problem1(-5)` is `-4` and `problem1(1)` evaluates to `-0.2642`.

2. A **compander** is a device that passes audio signals from -1V to 1V without change, but reduces the gain of higher magnitude signals, in part to prevent very large transient (short length) signals from saturating the system. You will learn more about companders in Communication courses.
   An example of a compander is the rule:

$$v_{out}(v_{in})=\begin{cases} \frac{1}{2}v_{in}-\frac{1}{2}, & -2\le v_{in}\le -1 \\ v_{in}, & -1<v_{in}<1 \\ \frac{1}{2}v_{in}+\frac{1}{2}, & 1\le v_{in}\le 2 \end{cases}$$

   Note that the horizontal axis (the "x" axis) is $v_{in}$ and the vertical axis (the "y" axis) is $v_{out}$. Write a MATLAB program that takes a vector of values for $v_{in}$ and returns the associated $v_{out}$ vector. It should have the following function definition line:

```
function vout = problem2(vin)
```

   Test the code by verifying that the command `problem2([-2 0 2])` evaluates to `[-1.5 0 1.5]`.

3. In the "Nested Loops" section on page 167, there is a code segment called "FindResistors" that outputs all combinations of 4 resistors taken 2 at a time. However, it does not recognize that R1 = 10, R2 = 22 is the same combina-

tion as R1 = 22, R2 = 10. Modify the code to avoid listing duplicates like these. Hint: change the start index of i2 from 1 to something else. This requires some thought; it may help to find the pattern by crossing out the repeats in the output table. It should have the following function definition line:

```
function problem3()
```

4. The "Nested Loops to Search for Best Solutions" section on page 171 has code to find the two standard-value resistors whose value in series is the nearest to a given desired value. The associated Practice Problem asks you to modify that to return the two standard-value resistors whose value in parallel is the nearest to a given desired value. Modify the code so that it returns the best single resistor, or set of two resistors in series or parallel, that is closest to the given value. The function definition is:

```
[R1best,R2best,result] = problem4(rDesired)
```

The `result` is the equivalent resistance computed with R1best and R2best, which shows both how close the match to rDesired is and whether R1best and R2best should be connected in series or parallel. If the closest match is a single resistor, then R2best should be 0.

Test the code by verifying in the command line that one can create an 8.5 ohm resistor by placing a 1 ohm and a 7.5 ohm resistor in series, equivalently

```
[r1, r2, result] = problem4(8.5)
```

evaluates to

```
r1 = 1.0, r2 = 7.5, result = 8.5
```

Similarly verify that the best way to create a 2.555 ohm resistor is by placing a 4.7 ohm and 5.6 ohm resistor in parallel, equivalently

```
[r1, r2, result] = problem4(2.555)
```

evaluates to

```
r1 = 4.7, r2 = 5.6, result = 2.5553
```

5. The following sum will calculate π in the limit as N grows large.

$$f(N)=\sum_{k=0}^{N}\frac{\sqrt{12}}{(-3)^k(2k+1)}$$

Calculate it by writing a function using the following function definition:

```
function result = problem5(N)
```

To test it, setting N=7 will return π to 4 decimal places of accuracy.

6. Use MATLAB to write a function that solves the math problem:

   ```
   result = (x + y) * 7
   ```

   The function definition is:

   ```
   function result = problem6(x,y)
   ```

   The result should be in base 2. The arguments will be in base 2 and should be provided with quotes. To check your work:

   ```
   problem6('101','110')
   ```

   should return:

   ```
   1001101
   ```

7. Fourier series analysis is an important technique from the signal processing subdiscipline of electrical engineering. It involves the fact that any periodic waveform can be broken up into a sum of sinusoids. It sounds impossible that a sharp-edged function like a square wave could be composed of a sum of smooth sinusoids (indeed, Fourier's original 1807 paper on the subject was denied publication for this reason) but in the limit of summing an infinite number of sinusoids, it is true. As an example, create a function with the following function definition

   ```
   function f = problem7()
   ```

   that calculates the following sum with t being a 1000 element time vector ranging from $0 \le t \le 20$ (f and t are bolded because they are vectors):

   $$\boldsymbol{f} = \sum_{k=1,3,5,\ldots}^{49} \frac{\sin(k\boldsymbol{t})}{k}$$

   In words, this means to first create a time vector **t**, and then initialize the result vector **f** to be filled with a vector of zeros that is the same size as the **t** vector. Next, create a `for` loop using loop variable k that starts at 1 and increments in steps of 2 so it becomes every odd number up to 49. For every iteration of the loop, add the vector sin(k **t**)/k to **f**. This is a vector because **t** is a vector. The function returns **f**, but check your work by embedding a plot(t, f) at the end of your code. It should look similar to a squarewave. As the upper limit of the sum, 49, is increased the function will appear progressively more similar to an ideal square wave.

8. Challenging!: You want to create a 72Ω resistor, but that is not a standard value. One way to build it is with a 33Ω and 39Ω in series. Another way is with a 75Ω and 1800Ω in parallel. Numerically, both these options work out to the same 72Ω resistor. If you do a Monte Carlo analysis of the two options with 10,000 simulations, do the two options have the same spread of tolerances if they all made with 5% resistors? For this problem, provide only the two histograms of data, labeled to show which is the series and which is the parallel solution. Also, state which is the better method or if both are equally preferable.

# 6 SPICE

## OBJECTIVES

After completing this chapter, you will be able to use LTspice to do the following:

- Use LTspice to create a schematic
- Analyze a DC circuit, and find the internal voltages and currents, as if with a DMM
- Analyze a time-varying circuit, and find internal voltages, as if with an oscilloscope
- Analyze a circuit whose output changes with frequency
- Describe the several meanings of "ground"
- Describe the differences between energy and power, and use LTspice to compute them
- Simulate two common types of ICs: the 555 timer and the operational amplifier

## CIRCUIT SIMULATION

General circuit simulators calculate the voltages and currents in arbitrary circuits. They have been in existence since the early 1960's, encouraged by increasingly complex circuit designs and initially funded by the US Department of Defense. An undergraduate class project in Berkeley spurred some of the first precursors to the open-source program we now know as SPICE, the Simulation Program with Integrated Circuit Emphasis. The first complete version was authored by Laurence Nagel in 1973, and SPICE has since evolved to become the *de facto* open-source standard for all circuit simulation packages.

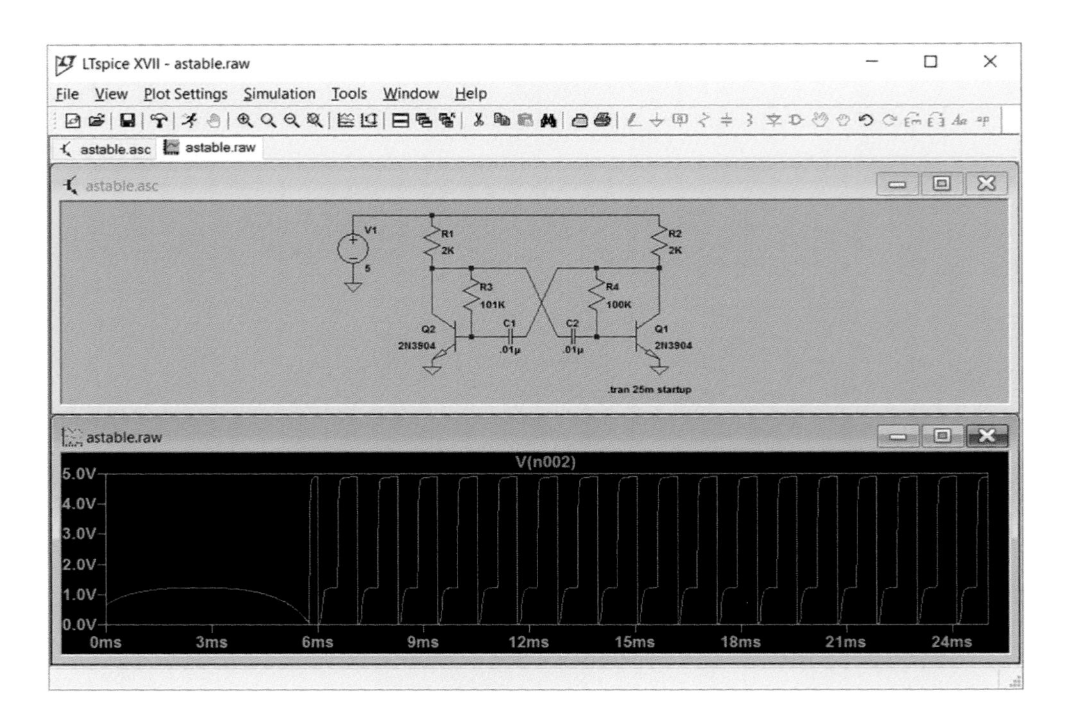


Most popular modern circuit simulators use SPICE at their core, including LTspice by Analog Devices, PSPICE by OrCAD, and Multisim by National Instruments. This text uses LTspice, which is free and available for both Windows and Mac operating systems.

## UTILITY

There are often two stages involved when solving real-world engineering problems: **design** and **validation**. In the design phase, the engineer uses the problem specification to create the schematic using techniques taught in circuit and logic design courses. The design phase cannot be aided by circuit simulation, although there are other packages that can assist with this phase, such as filter design programs.

The validation phase confirms that the circuit created in the design phase works as intended. Circuit simulators assist this phase by testing how the design performs in both ideal and non-ideal conditions, and they can evaluate performance given a range of component tolerances, temperatures, and operating conditions.

## INSTALLING LTspice

Download LTspice from the Analog Devices website (*http://www.analog.com*). Versions are available for both PC and Mac operating systems. Both are relatively small downloads, and despite possible differences in version numbers, both offer near-identical user experiences. Microsoft users should choose the 64 bit version if prompted during installation.

When the installation is complete, the program will automatically load and show the screen below:

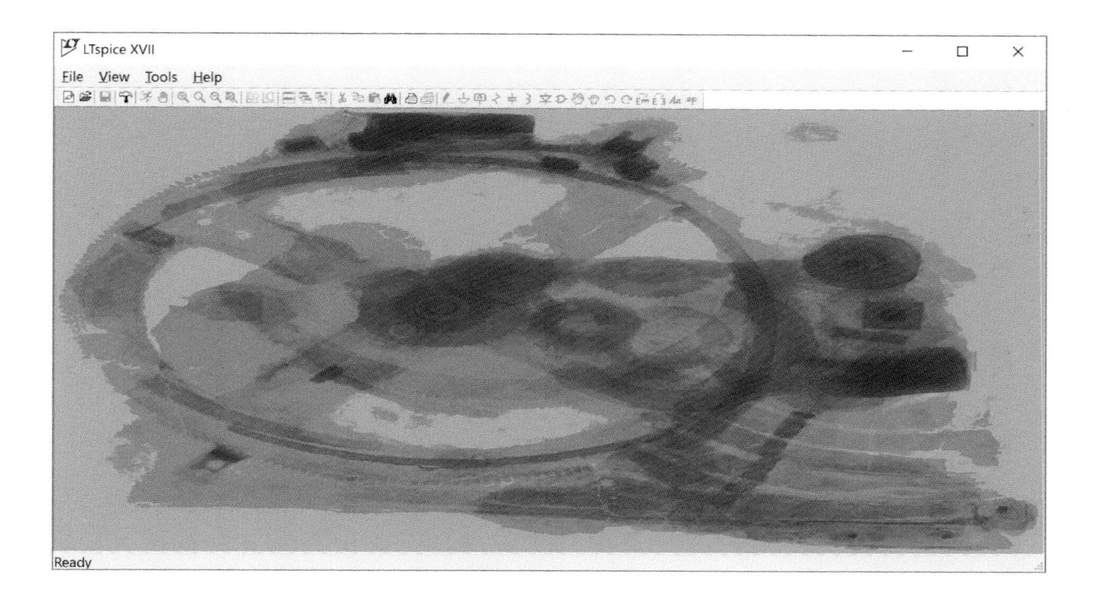


## STEPS IN SIMULATING

This textbook describes the following three steps in circuit simulation in detail:

1. Draw the schematic (i.e. schematic capture). Three different types of entities must be defined in this stage:
   A. Components, such as voltage and current sources, resistors, capacitors
   B. Ground
   C. Wires connecting the above elements

2. Define the analysis type. In this textbook, we will describe the three most common types:
   A. DC Analysis: DC analysis assumes all sources are constant, so the outputs will be constant.
   B. Transient Analysis: Transient analysis measures time-varying outputs.
   C. AC Analysis: AC analysis measures how the circuit responds to different frequencies of an input sinusoidal source.
3. View the results. Depending on the analysis type, this output may be a number, like a reading from a digital voltmeter; a time-dependent waveform like an oscilloscope trace; or a frequency-dependent Bode plot.

### *TECH TIP: WHAT IS GROUND?*

Unlike most components, there are three commonly used symbols for ground: ⏚ ▽ ⫻. LTSpice uses the triangular symbol shown in the center. To make matters more confusing, any of these ground symbols can mean any of three different concepts, depending on the context:

1. Ground can mean a physical connection to the earth, often through a copper spike driven into the ground. Ground is required by certain types of radio antennas, as well as by electrical power safety codes to protect the user in case insulation fails.
2. Ground can also refer to the set of conductors that return current to the power source in a circuit. For example, most electrical equipment in automobiles has two power connections: one connected by wires through a fuse to the +12V source, and the other to the metal chassis of the car, called "ground." The car chassis is connected to the negative side of the electrical power source by a short, thick ground wire.

3. In this chapter, however, ground refers to an imaginary, arbitrary voltage reference point in a circuit. Voltage is always measured between two points, which makes it awkward to describe in a circuit simulator, since the number of possible combinations of any two points in a circuit rises very rapidly with circuit complexity. When the user attaches the ground symbol to a point, say at "Point B," she is saying all voltages are measured relative to Point B. *A single ground symbol does not change the circuit behavior in any way*; it just lets us say "the voltage at Point A is 6V," rather than "the voltage between Point A and Point B is 6V." In other words, we define the voltage at ground to be 0V, which lets us measure all other circuit voltages relative to ground.

Ground is often placed at the most negative point in the circuit, which is often the negative terminal of the power source. In this case, all other voltages in the circuit (relative to ground) will be positive, and all current in the circuit will return through ground to the power source.

The schematic below (L) shows a number of voltages defined between various points. The adjacent schematic (R) adds a ground symbol, which lets the same voltages be re-defined in a much simpler manner. Which would you rather read?

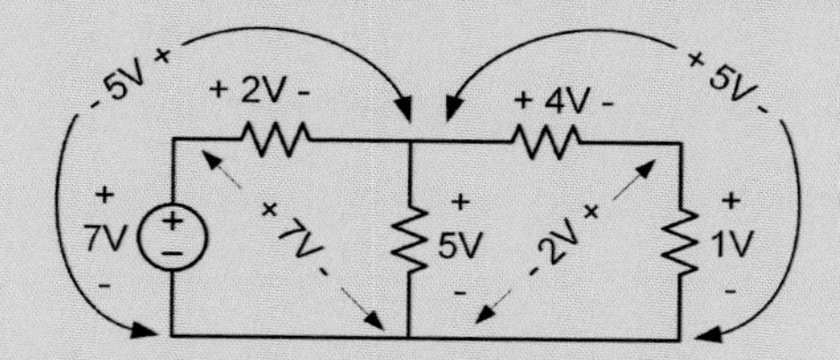

Voltages specified without ground

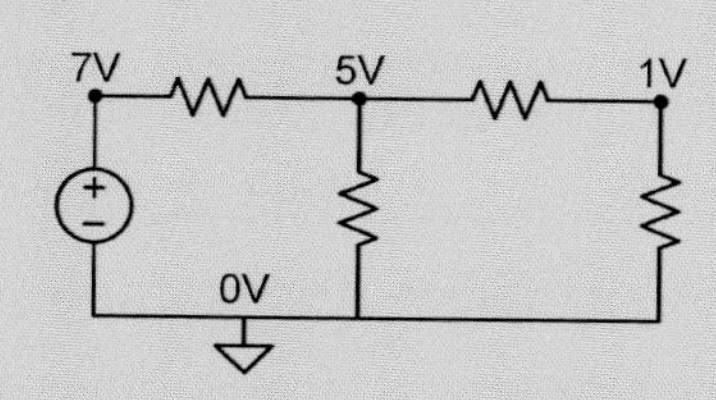

Voltages specified with ground

## PRACTICE PROBLEMS

1. The schematic below introduces a new component, the Bipolar Junction Transistor, or BJT. Its operation is covered in a future electronics course. You do not need to understand how the BJT works to be able to use the concept of ground to simplify the circuit. Voltages between various points are shown. If ground is placed at the bottom of the circuit, determine the voltages relative to ground at points A, B, and C in the circuit. Use the example in Tech Tip on the previous page as a guide, and remember that voltages do not change along a wire, only across components. ®

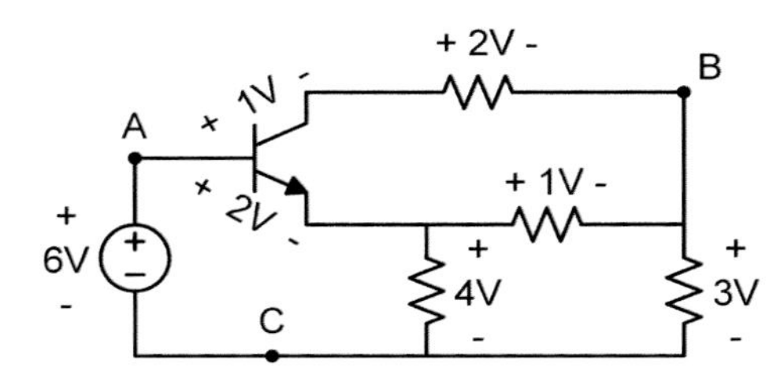


## SCHEMATIC EDITOR

Circuit simulation begins with **schematic capture**, which means creating a schematic diagram in the schematic editor. A schematic is composed of **components** (such as voltage and current sources, resistors, capacitors, inductors, diodes, transistors, and integrated circuits), at least one **ground** symbol, and **wires** connecting the components and ground. Every component has two pieces of associated data: its name, called its **reference designator**, and data describing its **value**. For example, a particular resistor may be named with a reference designator R11 and have a value of 4.3 kΩ, or a voltage source may be called V1 and have a value of 12V.

### *Starting a New Schematic*

The Windows application start button appears as follows: 

When started, LTspice periodically checks for new devices added to its library. Analog Devices provides this circuit simulator for free because its tight integration with the company's integrated circuits encourages engineers to use its devices. Since this textbook only uses basic components, it is safe to skip these library updates if you wish.

Once loaded, the program will appear as shown on page 191. The grayed-out logo in the center of the screen indicates that there is no current schematic. To create one, choose File → New Schematic from the menu. The logo will disappear, and the program's title bar will show "Draft1.asc" to indicate that a new schematic has been created with the default name. Change the name to "Schematic1" by using File → Save As and notice the change in the title bar text.

### *Selecting the Library*

The library is a collection of pre-defined components, including voltage and current sources, resistors, inductors, capacitors, and integrated circuits, all of which can be selected and drag-and-dropped to create a schematic. All of these components, and the libraries they are contained in, are selected using the button on the main toolbar. Choose it to display the Select Component Symbol dialog box, as shown below:

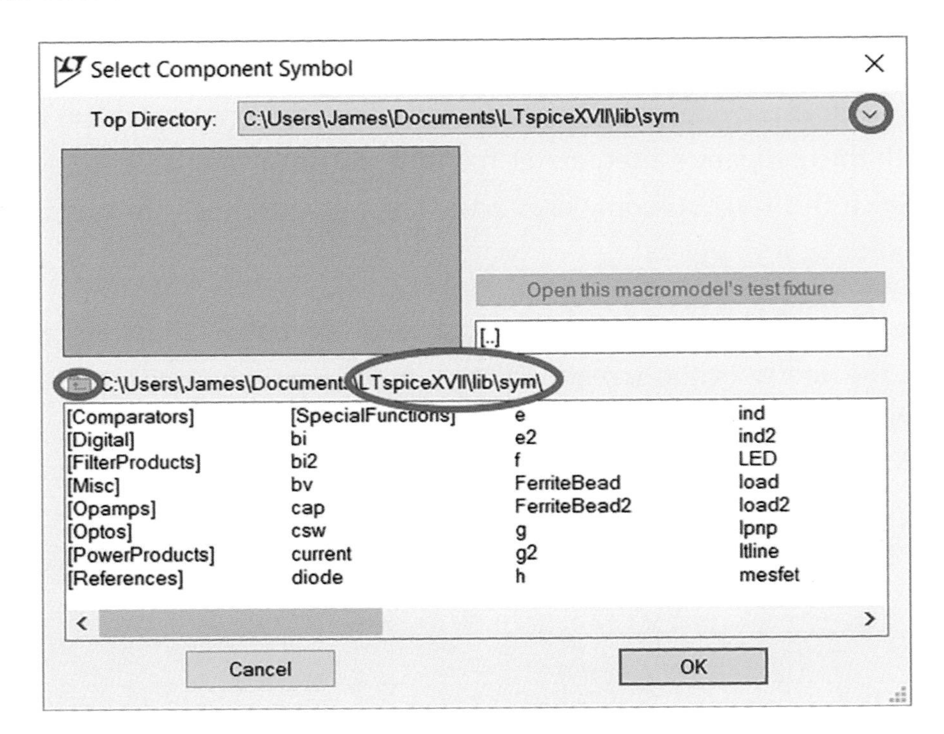


Note the folder in which the library is defined, highlighted with a red ellipse in the figure above. If the component dialog does not show components and component folders, beginning with the [Comparators] folder, it is because the library folder is pointing to the wrong location. Change it to the C:\Users\YourUserName\Documents\LTspiceXVII\lib\sym folder using the Top Directory dropdown selector and the Move Up Folder selector highlighted above.

***Defining Voltage, Current Sources***

To create a voltage source, select the component option using the button on the main toolbar. The component dialog box should appear as shown above, still set to the correct library location selected in the above step. Move the horizontal scrollbar until "voltage" is highlighted. Select it, click to place it in the schematic, and press the Escape key to escape from voltage source placing mode.

To select the voltage level of the source, right-click the symbol (*not* the text next to the symbol) and enter the desired voltage in the "DC value (V)" textbox. Leave the "Series Resistance (Ω)" textbox blank. Current sources are set in the same manner using the "current" component.

The LTspice voltage and current sources are very flexible; they can represent constant voltage or current sources, a squarewave, a sinusoidal wave, or a number of other commonly-encountered waveform types. By default, it is a DC (constant) source. To select a different waveform type, right-click the symbol (*not* the text) and choose the "Advanced" button.

To change the designator ID of the source, right-click the text surrounding the source. Typical names for voltage sources are V1, V2, etc., and I1, I2, etc., for current sources.

### *Other Common Components: Resistors, Capacitors, Inductors, Diodes, Transistors*

Resistors, capacitors, inductors, diodes, LEDs, transistors, and more can be selected using the component dialog box. The only abbreviated names are "res," "cap," and "ind" for resistor, capacitor, and inductor. These three and the diode have dedicated toolbar buttons, also: . You can use the prefix symbols M, k, m, and u to stand for Mega, kilo, milli, and micro, respectively (note that letter u is used instead of μ). For example, it is less error-prone to specify a 2.2μF capacitor as 2.2u rather than 0.0000022.

### *Manipulating Component Placement*

Components can be rotated in 90° increments before they are placed using the Ctrl-R key combination. Once they are placed, they can be rotated or moved by selecting the tool, then selecting the component, and then either moving it with the mouse or pressing Ctrl-R. Components can be deleted by selecting the tool and then clicking on the component to be deleted. Mistakes can be undone with the undo tool ; however, the undo tool is not mapped to the typical Ctrl-Z shortcut.

### *Wire, Ground*

Wires are drawn after selecting the wire tool, . Unwired connection points on components appear as open squares; once connected with a wire, the square disappears. The ground symbol is placed after selecting the ground tool, . Make certain the open square on top of the ground disappears after connecting it to the rest of the schematic with a wire. If this is not done, or if the ground is not placed, the simulation will fail and report the warning, "This circuit does not have a conduction path to ground."

### *Operational Amplifiers and Other Integrated Circuits*

Future courses will likely introduce myriad new components, called integrated circuits (the "IC" in SPICE). These can be found from the component dialog box by opening up the folder of the IC category, and they include components such as comparators and opamps. Sometimes generic components will be available; an example of this is the "opamp" available in the "Opamps" library, which is meant to represent a generic, somewhat-idealized opamp.

### *Summary of Schematic Capture*

- Schematic capture is the first step of circuit simulation.
- Components: created using  toolbar button.
  - Examples: voltage source, resistors, inductors, transistors, ICs.
  - Rotate them before placing using Ctrl-R.
  - Reference Designator: Click on the *text* to change the component name, (e.g. C2).
  - Value: Click on the *symbol* to change the value (e.g. 4.7uF).

  Ground: created using the  toolbar button.
  - At least one required per circuit.
  - Often placed at most negative part of the circuit.
  - Voltages measured relative to this point.
- Wires: created using the  toolbar button.
  - Open squares at connection points disappear when connected with a wire.

### PRACTICE PROBLEMS

2. Create a schematic of the following circuit, print it, and annotate it with your name. Save it to your work folder to be used in the next Practice Problem. Using the Windows Snipping Tool or the Mac Grab Tool, cut and paste the image into a Microsoft Word file. ®

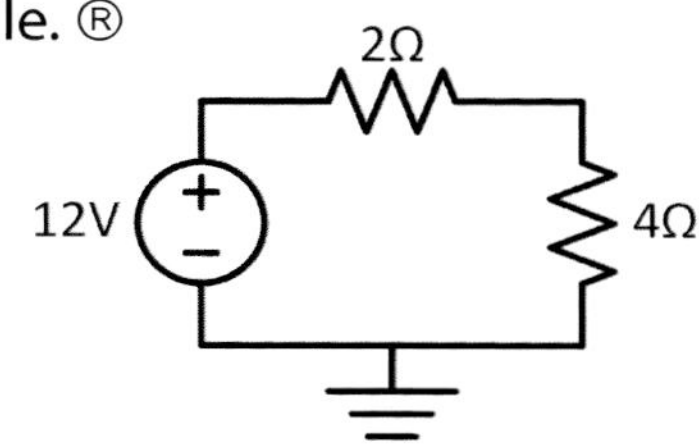


## TYPES OF LTSPICE SIMULATIONS

There are three commonly-used SPICE analysis types at the undergraduate level: DC Operating Point, Transient, and AC Sweep.

***DC Operating Point:*** this is a complicated name for a simple concept. Use this analysis type if the circuit has only unchanging sources, meaning constant-voltage and constant-current sources. In this case, all voltages and currents in the circuit will also be constant, and would be measured in a real-world circuit using a digital multimeter. There are no options for this simulation, and the results are viewed by hovering the cursor over the wire or component of interest.

***Transient Analysis:*** This is the analysis type to use if the circuit has changing voltages and currents. Examples include circuits whose sources step from one voltage to another, or are defined to be a square wave, or if the circuit oscillates. The real-world circuit would require an oscilloscope to view how the circuit voltages and currents change in time; the simulated results are viewed by clicking the wire or component of interest after simulation to show their voltages on a simulated oscilloscope.

***AC Sweep:*** Electrical engineers are often concerned with how a circuit changes a purely sinusoidal source of various frequencies. For instance, a high-fidelity audio amplifier should exhibit the same signal gain for input frequencies spanning the range of human hearing (often taken to be 20Hz – 20kHz), or a lowpass filter that routes signal energy to bass speakers should pass frequencies below about 150Hz but attenuate higher-frequency signals. This analysis is performed with an AC Sweep, in which one source is defined to be the source of the variable-frequency AC signal, and the resulting change in the signal magnitude and phase is displayed vs. frequency. This is the equivalent of attaching a spectrum analyzer to a real-world circuit, and the resulting magnitude vs. frequency plot is commonly called a Bode plot.

## DC SIMULATION WALK-THROUGH

The second and third phases of circuit simulation, defining the analysis type and viewing the results, is best taught by example. The first of the three main types of circuit analysis this chapter covers is DC simulation, and it will be taught by completing an analysis of the circuit shown below:

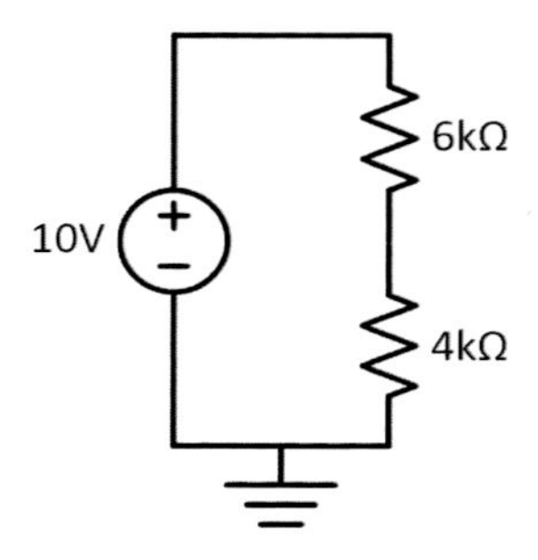


Follow these steps to gain a solid working knowledge of DC circuit analysis:

***Open up a blank schematic***

- Open LTspice and from the menu, File → New Schematic
- Save it immediately using menu, File → Save As. Give the schematic a simple name with no spaces under "Name" (e.g. ComputerTools_DC). Save it in your personal data directory for this course.

***Add a 10V source***

- Click the component icon. Find the "voltage" component—typing the first few letters will scroll to it—double-click to select it, and place it in your schematic by clicking once. If you click again, you'll place a second voltage source; to prevent this and get out of "create voltage source" mode, press the Esc key.
- How to delete and move components: rather than select-and-drag, or select-and-delete, as in most programs, select the action first, and then select the thing you want to change. Create a second voltage source now using the above steps to practice:
  – To delete the second voltage source, select the delete tool, and then click on the offending voltage source. Press Esc to exit delete mode.
  – Move your desired V1 source to the left middle screen by selecting the move tool. If you forget which icon it is, hover the mouse over the icons for a tooltip. Select the voltage source you want to move, then move your mouse and click to place it. Press Esc to exit move mode.
- To set the voltage to 10V, right-click inside the symbol. Do not right-click on text outside the symbol or you will change the **name** of the source but not change its **value**.
  – Set the DC value to 10V (the "V" at the end is optional), leave the series resistance box empty and press OK.
- Save your work! Ctrl-S or from the menu, File → Save.
- ***Review:*** to make a 10V source, select the part using and choose "Voltage", then set the value of the source by right-clicking inside the symbol.

***Add two resistors***

- You could click the and choose "resistor," but resistors are so common they have their own icon:. Click it and place two resistors stacked vertically but separated by a small distance to the right of the voltage source. Click "Esc" to exit resistor-placing mode.
- Don't like the orientation of your resistors? Try placing one horizontally-oriented. To do this, choose the resistor icon again as you did above,

but before placing the resistor, press Ctrl-R to rotate it. Place this horizontal resistor, then delete it using the technique you learned in the previous section.

- Right-click resistance symbol (*not* the text) and make the top one 6k and the bottom 4k. Save your work.
- ***Review:*** To place a resistor, choose ⩽, and right-click the symbol to set the resistor's value. Press Ctrl-R before placing to rotate.

***Add the ground and wire everything together***

- Click the ⏚ icon and place it below the voltage source so there is a little space separating them.
- Select the wire tool ✎ to wire the parts together.
- Click on one component end, and then click on another component that you wish to wire together. The open square ends of the wire will vanish, and the wire will stop following the cursor. Clicking intermediate points while wiring will allow you to create corners.
- Mistakes can be deleted using the ✂ tool.
- Repeat to wire all components together. When you place a wire on another wire, it will form a closed square indicating a connection. Compare your result to the figure on the right. Save your work. Hint: To check your work, check to make sure there are no open boxes (unwired connection points).

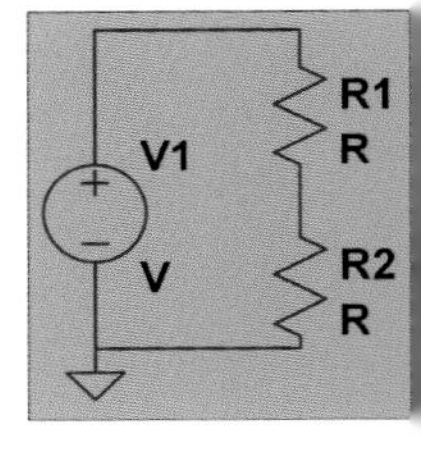


- ***Review:*** Place a ground using ⏚ and wire components using ✎.

***Run the analysis***

- Specify the type of analysis to run. From the menu, choose Simulate → Edit Simulation Cmd and choose the "DC op pnt" tab (DC operating point) to determine all voltages and currents for this example with constant sources. Choose "OK" and place the ".op" text that appears anywhere on a blank area of the schematic.
- Select Simulate → Run from the window or choose the 🏃 button. An analysis window will pop up and show the voltages at, and currents through, various cryptically-named nets like "n001".

- To determine the voltage along any wire relative to ground, hover the cursor above the wire and look in the status bar at the bottom of the window. This duplicates probing the wire with a digital multimeter.
- To determine the current flowing through any component and the power that it dissipates, hover the cursor over the component and look in the status bar.
- ***Review:*** Find DC voltages and currents using Simulate → Edit Simulation Cmd and choosing "DC op pnt". Run the simulation using the button. View voltages and currents by hovering over them in the schematic window.

*PRACTICE PROBLEMS*

3. Simulate the following circuit you built in Practice Problem 2, and find the voltage across the 4Ω resistor. ®

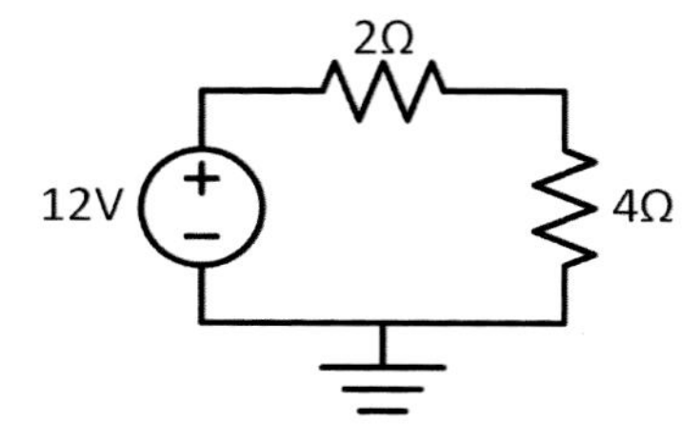

## *TECH TIP: POWER VS. ENERGY (WORK)*

Many people use the terms "power" and "energy" interchangeably, but these terms have very different meanings. Energy, measured in Joules (J), is the ability to do work—examples of a quantity of energy include a coin lifted by 1 foot (20J), two AA batteries (20kJ), a ham sandwich (2MJ), or 20 gallons of gasoline (2.4GJ). The rate at which energy is used is power, measured in Watts (W). A Ferrari will be able to deliver more power than a Volkswagen Beetle, even though both may start with their gas tanks identically full (i.e. they begin with the same energy, but the greater power of the Ferrari will cause it to use that energy more quickly). A car going up a hill will use more power than the same car traveling on flat ground. The ham sandwich will power a young child for longer than an adult, and the AA battery will light a 150mW LED for about twice as long as a brighter LED rated for 300mW of power consumption.

Mathematically, energy is the time-integral of power (or power x time, if power use is constant), and power is the time-derivative of energy (or energy / time, if power use is constant).

$$energy = \int power(t)dt = power \times time \ \ (\text{if power is constant})$$

$$power = \frac{d}{dt} energy(t) = \frac{energy}{time} \ \ (\text{if power is constant})$$

## PRACTICE PROBLEMS

4. Use LTspice to determine how much power is being delivered by the voltage source in the figure below:

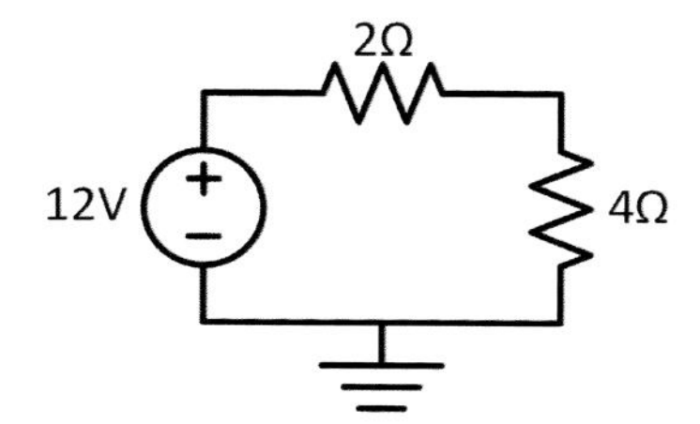


5. How long will 8 1.5V AA batteries power the circuit above if a single AA battery holds about 10kJ of energy? Hint: solve the $power = \frac{energy}{time}$ equation given above for time). ©®

### PRO TIP: ETHICS II

Your engineering integrity will likely be tested many times throughout your career, which is why the Accreditation Board for Engineering and Technology (ABET) requires that ethics be an integral part of engineering education. Besides the IEEE Code of Ethics, described in an earlier Pro Tip, engineers who earn the Professional Engineer (PE) certification promise to abide by a similar set of rules set by the National Society of Professional Engineers (NSPE). This organization represents all engineering fields, including electrical, computer, mechanical, civil, biomedical, chemical, nuclear, and petroleum.

The preamble of the NSPE Code of Conduct for Engineers (2007) states:

> *Engineers shall at all times recognize that their primary obligation is to protect the safety, health, property, and welfare of the public. If their professional judgment is overruled under circumstances where the safety, health, property, or welfare of the public are endangered, they shall notify their employer or client and such other authority as may be appropriate.*

David Jonassen developed a generic framework to help you use this NSPE Code of Ethics, or the IEEE Code of Ethics, with any engineering ethics problem (*Journal of Education Education*, vol. 98, no. 3, 2009). His analysis method consists of five steps:

- State the problem. Define the ethical problem clearly in a few words.
- Get the facts.
- Identify and defend competing moral viewpoints. Most ethical problems involve multiple perspectives, and each viewpoint may lead to a different conclusion.
- Identify the best course of action.
- Qualify the course of action with facts.

When faced with a difficult decision, consider documenting your decision using this framework.

## *TECH TIP: 555 TIMER*

There are over ½ million integrated circuits (ICs) sold just by Digi-Key, one of the largest electrical parts vendors. Of these, the 555 timer is historically the most popular. Since its introduction by Signetics in 1972 this chip has outsold all other types of ICs. It has been used in medical monitors, spacecraft navigation systems, and children's toys. Although it can be used many different configurations, one of the most common is as an **astable multivibrator**—in simpler words, an **oscillator**. The frequency is set using two resistors and a capacitor; it does not change significantly with changes in the powering voltage. The schematic using it as an astable multivibrator and the pin numbering for the IC are shown below.

Notice that although the schematic symbol is rectangular like the IC, the pin numbering of the schematic does not have to correspond to the pin positions on the actual device, which allows the schematic to be drawn more simply. For instance, the schematic shows pins 4 and 8 next to each other since they are electrically connected with a wire, although on the physical device those pins are diagonally across the body of the IC.

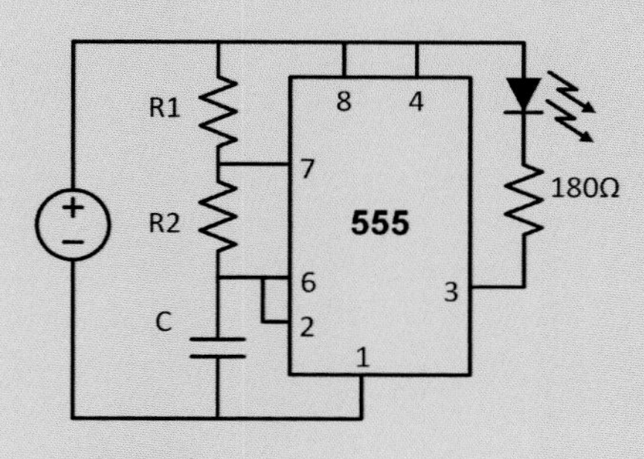


**Schematic**

The design equations for this circuit are:

LED blinks on for 0.693*R2*C seconds

LED blinks off for 0.693*(R1+R2)*C seconds

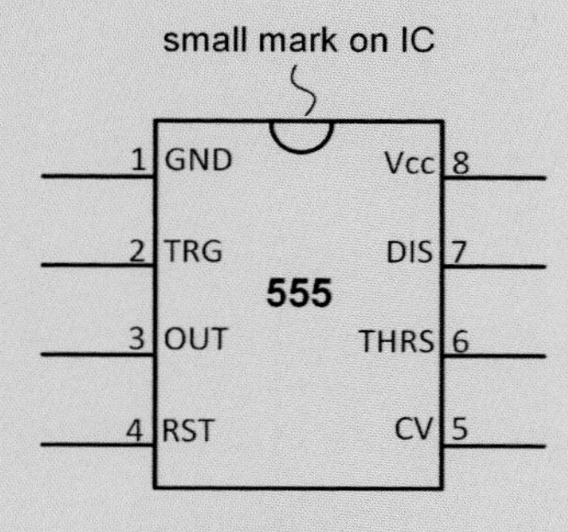


**Pin numbering**

## PRACTICE PROBLEMS

6. Use MATLAB to find the values for R1 and R2 so that the LED is on for 1/3 second and off for 2/3 second. Use a C = 1μF capacitor. ®

## TRANSIENT SIMULATION

The second of the three types of circuit analysis this chapter describes is transient analysis. It will be taught by completing an analysis of the circuit shown below (L). Note that although similar, this is not the solution to the Practice Problem. Follow through the steps that follow to gain a solid working knowledge of transient circuit analysis. When the schematic capture is complete, it should look as shown below (R). LTspice fixes the location of the IC pins in the same positions as on the physical chip. Although this makes wiring up a physical circuit from an LTspice schematic easier, it makes the schematic more difficult to read.

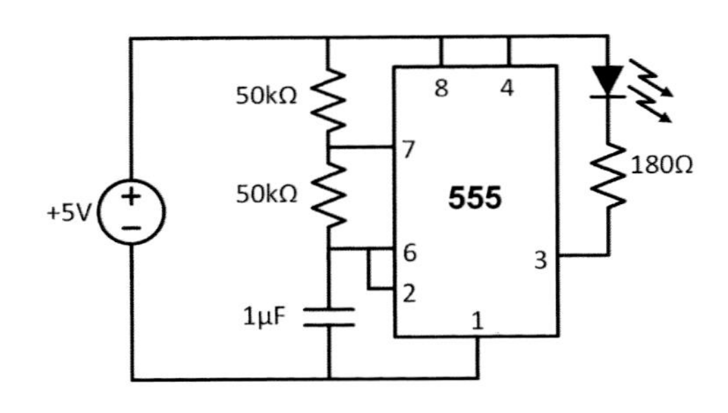


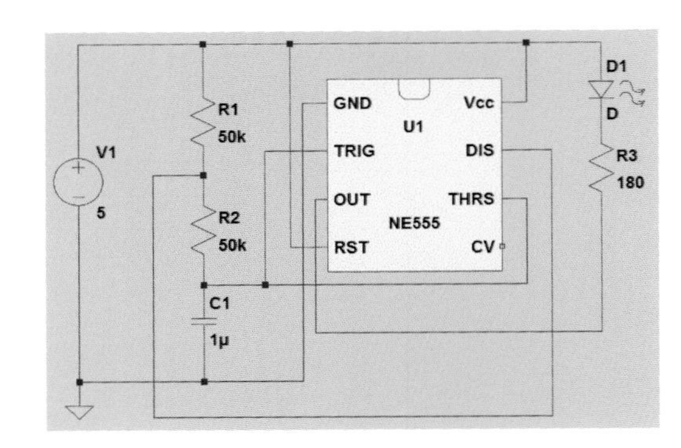


***Open up a blank schematic***

- Open LTspice, and from the menu, File → New Schematic
- Save it immediately using menu, File → Save As. Give the schematic a simple name with no spaces under "Name" (e.g. ComputerTools_Transient). Save it in your personal data directory for this course.

***Add voltage source***

- Click the component icon. Find the "voltage" component—typing the first few letters will scroll to it—double-click to select it, and place it in your schematic by clicking once. Right-click the symbol, not the text, and make it 5V.

***Add 555 integrated circuit***

- Integrated circuits are organized into folders. Choose the [misc] folder and select the NE555 component. The NE prefix is unimportant; it indicates a variant improving upon the original 555 design. Leave plenty of room between it and the voltage source to place the resistors and capacitor. You may need to alter the zoom with the tools that zoom in, pan, zoom out, and zoom to fit the entire schematic on the visible page, respectively.

***Add resistors, capacitor, LED, ground***

- Click the resistor icon, and Ctrl-R if desired to rotate, place, and right-click the component symbol (not text) and give it a value of 50k. Spice understands "k" and there is no need to add the Ω symbol. Do the same for the capacitor using the symbol and using 1u. Recall Spice uses "u" for μ. The LED is found under the main folder, and like the 555 IC has no parameters to set. The ground is placed with the symbol.

***Wire it up***

- Use the tool to wire the circuit together.

***Run the transient simulation analysis***

- Specify the type of analysis to run. From the menu, choose Simulate → Edit Simulation Cmd and choose the default "transient" tab to determine all voltages and currents for this example since the oscillator's output varies with time. To analyze the circuit from 0 to 0.5 seconds, specify 0.5 as the stop time, 0 seconds as the time to start saving data, and leave the rest of the options blank.

## DIGGING DEEPER

Curious about all of the other options for the transient simulation analysis? Most of the options are rather arcane and can be investigated under Help > LTspice > Transient Analysis Options. The only three commonly-adjusted options are *Stop time*, *Time to start saving data*, and *Maximum timestep*. *Stop time* is the duration of the simulation in seconds. Set a *Time to start saving data* value to greater than 0 seconds if early circuit behavior is not of interest. LTspice automatically chooses how frequently to compute output values—more often when output values are quickly changing, and less-frequently otherwise. This can be manually controlled by setting the *maximum timestep* value, often in milliseconds or less.

- Choose "OK" and place the ".tran 0.5" text on a blank area of the schematic.
- Select Simulate → Run from the window or choose the button. The following split schematic/analysis window will appear.

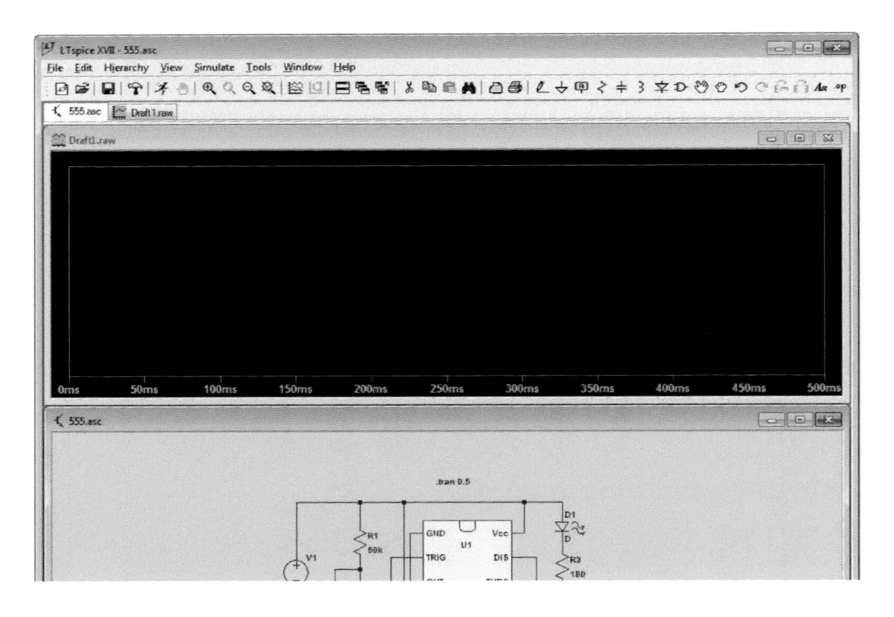


- Position the cursor over a wire in the schematic in which you want to see an oscilloscope-style plot of the voltage vs. time. In the example above, the cursor is hovering over the wire that is attached to pin 3, OUT. The cursor changes to resemble a red oscilloscope probe.
- Click on the probe to display the corresponding voltage waveform.

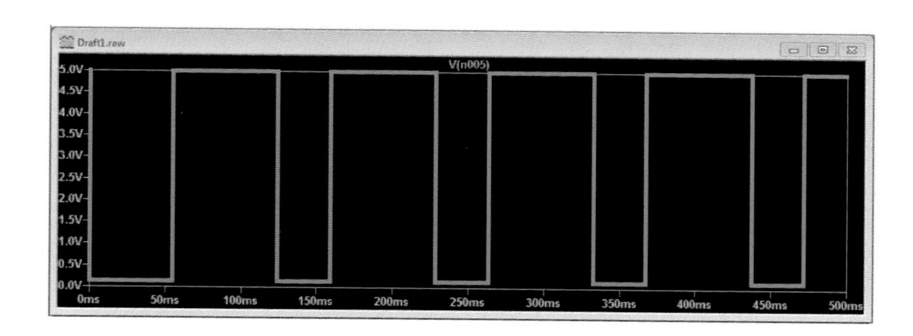


- Click on other wires to add traces. Right click on a trace and from the popup menu choose Edit → Delete to remove.

## *PRACTICE PROBLEMS*

7. Modify the above walkthrough to print one second's worth of data. Record a screen grab of the analysis window using the Snipping Tool (if on Windows) or Command+Shift+4 (if on a Mac). ®

## *TECH TIP: OPERATIONAL AMPLIFIERS*

The operational amplifier, or more simply "opamp" is a very commonly-used integrated circuit. It is available in a stunning number of varieties; Digi-key sells 35,000 different types of opamps. Opamps are thoroughly described in circuits courses; this textbook considers one application of an idealized opamp whose schematic symbol is:

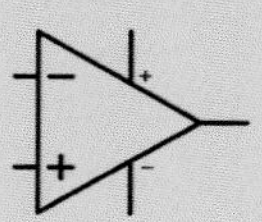

The opamp is used for many different purposes including amplifying, filtering, and even as an analog computer, integrating and differentiating input signals. An example of a lowpass filter is shown below. The signal input $v_{in}(t)$ to the opamp is on the left, the output is on the right, and the vertical wires in the middle of the opamp with the small polarity symbols are the DC power connections to the opamp.

The circuit shown below, known as a second order Butterworth lowpass filter, will pass input sinusoids with frequencies below a preset frequency $f_0$, but attenuate sinusoids of higher frequencies. It could be placed before a bass amplifier to ensure no high frequencies are sent to the woofer speakers.

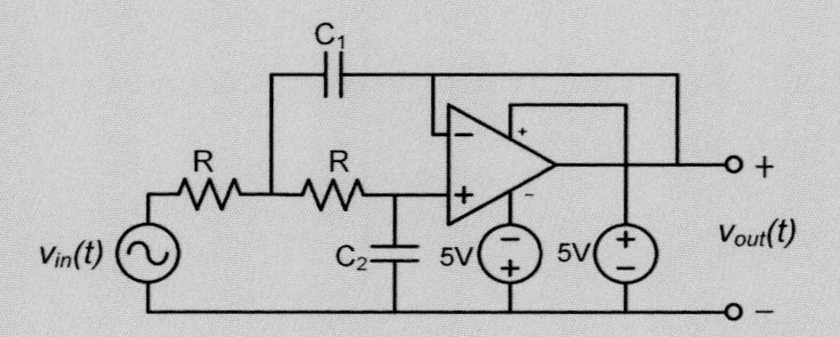


The right-most 5V source's wire crosses, but does not connect to, the output of the opamp. The convention in EE is that two wires crossing perpendicularly do not connect unless there is a large black dot drawn at their intersection.

The design equations for this circuit, given desired cutoff frequency $f_0$ and resistor value R, are:

$$C_1 = \frac{\sqrt{2}}{2\pi R f_0} \quad \text{and} \quad C_2 = \frac{C_1}{2}$$

## PRACTICE PROBLEMS

8. Use the above equations to select $C_1$ and $C_2$ to design a lowpass filter with a cutoff of 200Hz, a realistic value for a subwoofer speaker amplifier. Use 10kΩ resistors. ®

## AC SWEEP SIMULATION

The last of the three types of circuit analysis this chapter describes is AC sweep analysis, and it will be taught by completing an analysis of the circuit shown below on the left (note that although similar, this is not the solution to the Practice Problem). Follow the steps below to gain a solid working knowledge of AC sweep analysis. When the schematic capture is complete, it should look as below on the right. Notice that because ground is defined, there is no need to bring the lower wire out from the schematic to define $V_{out}$ between two points; it is now the single point to the right of the opamp.

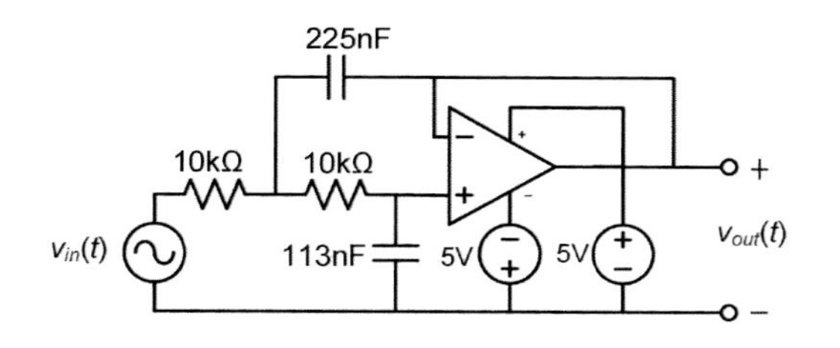


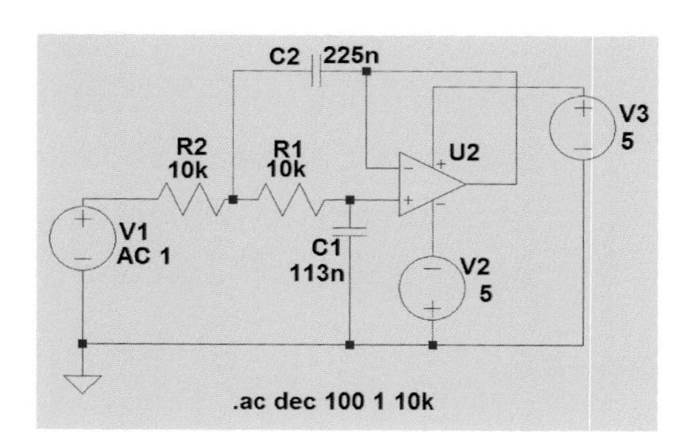


***Open up a blank schematic***

- Open LTspice, and from the menu, File → New Schematic
- Save it immediately using menu, File → Save As. Give the schematic a simple name with no spaces under "Name" (e.g. ComputerTools_AC). Save it in your personal data directory for this course.

***Add opamp integrated circuit***

- Integrated circuits are separated by folders. Choose the [Opamp] folder and select the "UniversalOpamp2" component. Typing the first few letters of the name will autoscroll to the right location. If you were physically prototyping the circuit you could achieve a more accurate simulation by selecting an opamp part number rather than the generic UniversalOpamp2 part.

***Add voltage sources***

- Click the component icon. Find the "voltage" component—typing the first few letters will scroll to it—double-click to select it, and place it in your schematic by clicking once. You may want to flip its direction using Ctrl-R. Right-click the symbol, not the text, to set the opamp power sources to 5V. You may need to zoom out with the tool to place all of the sources. To indicate that the input on the left is the AC sinusoidal source whose frequency will be swept during the analysis stage, leave the DC value blank, choose "Advanced" and then in the "Small signal AC analysis" panel enter "1" for AC Amplitude.

***Add resistors, capacitor, and ground***

- Click the resistor icon, place two resistors, right-click the component symbol (not text), and give them values of 10k. Spice understands "k" and there is no need to add the Ω symbol. Do the same for the capacitors using the symbol (you may want to rotate one using Ctrl-R before placing). The ground is placed with the symbol.

***Wire it up***

- Use the tool to wire the circuit together.

***Run the AC sweep analysis***

- Specify the type of analysis to run. From the menu, choose Simulate → Edit Simulation Cmd and choose the "AC Analysis" tab. AC Analysis determines how the magnitude of the output voltage changes as a designated input sinusoid varies in frequency. To analyze the circuit over a range of frequencies from 1Hz to 10kHz, choose the "Decade" type of sweep (this is the standard Bode-style range of logarithmic frequencies earlier studied), 100 points per decade (higher is more accurate, lower is faster), 1 as the start frequency and 10k as the stop.

- Choose "OK" and place the ".ac dec 100 1 10k" text anywhere on a blank area of the schematic.
- Select Simulate → Run from the window or choose the button.

The following split schematic/analysis window will appear:

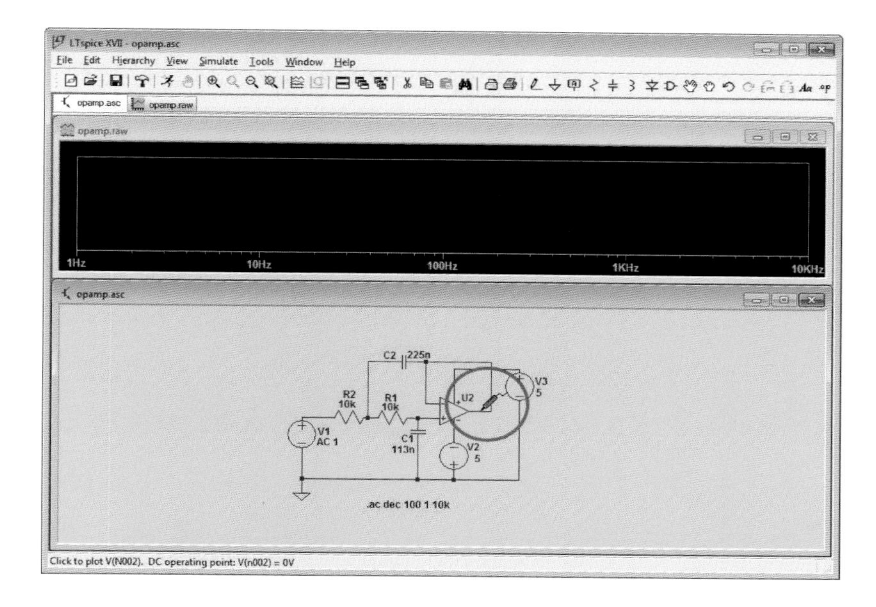


- Position the cursor over a schematic wire of whose voltage you want to examine an Bode-style plot of the voltage vs. frequency. The above example highlights the cursor hovering over the wire that is exiting out of the opamp. The cursor changes to resemble a red probe.
- Click on the probe to display the corresponding voltage.

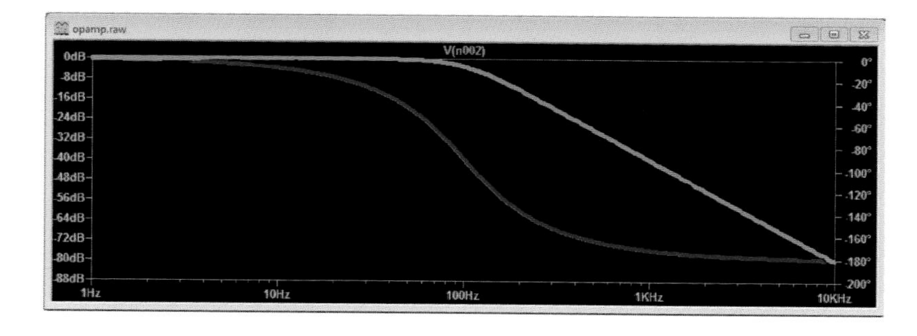

- Two plots are automatically created; one shows the magnitude of the signal, whose value is plotted relative to the axis on the left, and the other plot shows the phase, whose value is plotted relative to the axis on the right. These are Bode plots. To remove the phase trace, click on the right axis and choose "Don't plot phase." Notice how the data shows low frequency signals under about 100Hz are not attenuated. Earlier chapters described how dB are calculated using a logarithmic scale of $10^{dB/20}$, so $0dB = 10d^{0/20} = 10^0 = 1 = 100\%$ passed at low frequencies. Higher-frequency signals are reduced; the approximately -40dB of attenuation at 1kHz corresponds to the circuit passing just $10^{-40/20} = 10^{-2} = 1\%$ of the input signal.

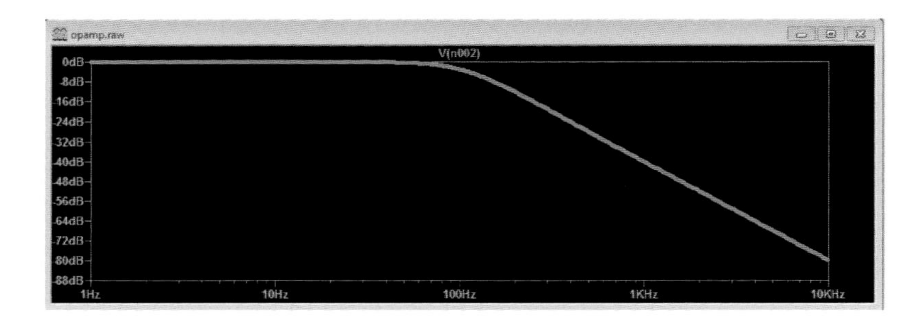


## *PRACTICE PROBLEMS*

9. Modify above walkthrough to use 20k resistors, but the same value of capacitors. Analyze the performance of the filter over the same 1Hz to 10kHz range. Record a screen grab of the analysis window using the Snipping Tool using Windows or Command+Shift+4 on a Mac. Does the cutoff frequency increase or decrease as resistor values are increased? ®

## ADVANCED TIP: USING NETS

The opamp schematic is functional, but looks messy because of the 5V sources providing power to the opamps. If we defined a symbol like +5V to be 5V and another -5V to be -5V, then we could significantly neaten the schematic while making the intent clearer. These symbols are examples of **nets**, as illustrated in the following schematic:

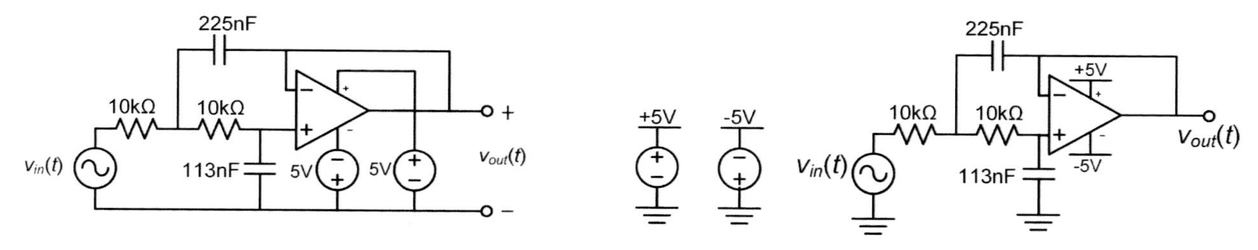


The circuit on the right is functionally the same as the previously-analyzed circuit on the left. The two voltage sources that are apparently unconnected define two nets called +5V and 5V, which are used to power the opamp on the right. It also uses multiple ground net symbols to simplify the schematic further. Practicing engineers much prefer this to the schematic on the left.

LTspice permits the definition of nets using the tool. As an example of using it to simplify the previous schematic, first create the following schematic capture:

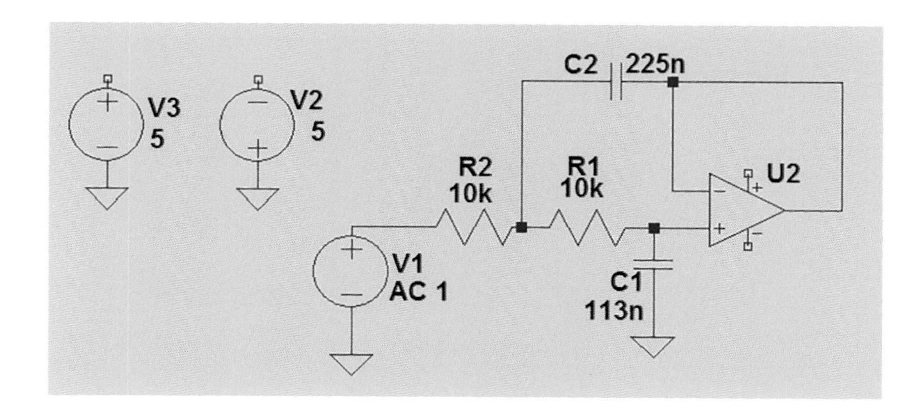

Then click the [icon] tool and create the net name +5V, of Port Type "none", and place as shown below. Similarly, create a -5V net name and place as shown in the figure below:

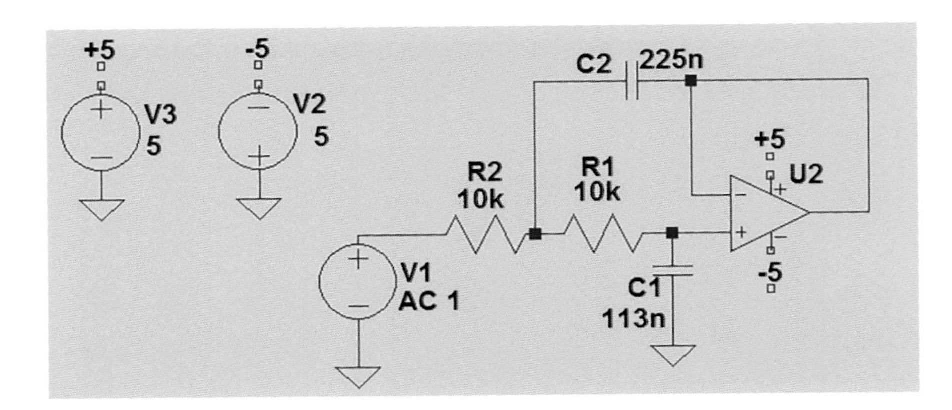


Wire them up as shown below. Now every time a +5V net symbol is used, it is functionally identical to being wired to a +5V voltage source whose more negative end is wired to ground. This is perhaps only marginally neater in this single-opamp example, but becomes far neater in more complex circuits that use multiple opamps.

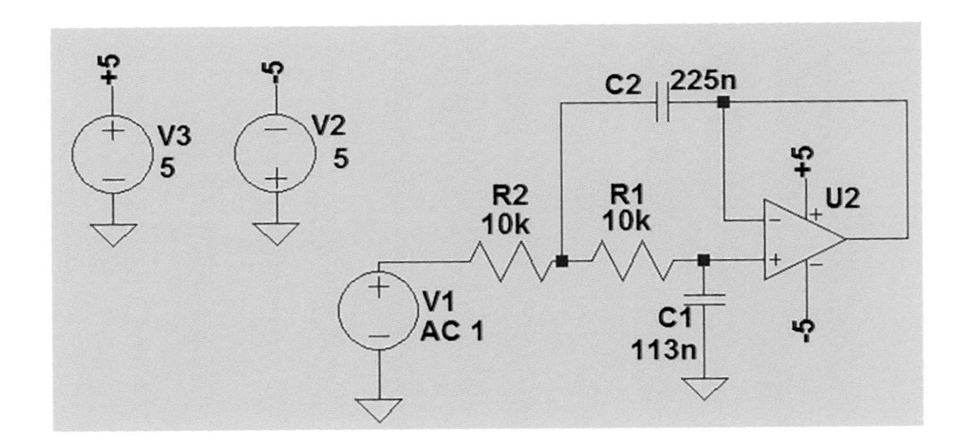

*PRO TIP: LIFELONG LEARNING*

In 1962, economist Fritz Machlup coined the phrase "Half-life of knowledge" to quantify how quickly knowledge becomes obsolete. The outlook is not sanguine for electrical engineers; about a century ago, it was estimated to be 35 years, while a half-century ago, this dropped to a decade, and in 2002, the president of the National Academy of Engineering stated it was between 2.5 and 7 years, depending on the subfield. Entire career fields are similarly mobile; a generation ago, an engineer could expect to carve out a niche in one well-defined area for one company—missile guidance systems by Lockheed, for example—and remain there for a career. No longer.

What can electrical students do to safeguard hard-won technology skills?

1. Accept that the skills learned as an undergraduate can be generalized and are designed to be transferable. MATLAB and Spice have existed for a long time (since 1984 and 1973 respectively), but both may be superseded by new technologies in a career lifetime. It is unlikely, however, that the underlying programming and circuit analysis skills gained by learning them will be similarly replaced. It is much, much easier to learn a second programming language or circuit analysis system after the first.
2. Continue to learn and retrain throughout one's career. Use IEEE membership to keep abreast of new software and hardware developments through section

meetings, and read publications like the monthly *IEEE Spectrum* that comes with IEEE membership. Many companies offer in-house training programs and reimburse educational expenses from weekend seminars to full university degrees. Most employers reimburse continued education training expenses required to maintain PE licenses. Electrical engineering also has a large presence in the relatively recent introduction of Massive Open Online Courses (MOOCS); some excellent ones are offered at *http://courses.edx.org* and through the IEEE.

Ultimately, it is the same love of learning that drives people to become engineers that keeps them current throughout their careers. Although the profession is changing more rapidly now than ever before, engineering culture is equally quickly adapting to the reality that education is not a college degree but a lifetime process.

## COMMAND REVIEW

### *Schematic Capture Commands*

File → New Schematic — Begin a new schematic diagram.

| | |
|---|---|
| (component icon) | Place a component |
| (component icon), Voltage | Place a voltage source (after selecting ) |
| (resistor icon) | Place a resistor |
| (capacitor icon) | Place a capacitor |
| (inductor icon) | Place an inductor |
| (wire icon) | Place a wire |
| (ground icon) | Place ground |

### *Zoom Tools*

(zoom icons) — Zoom in, pan, zoom out, and zoom to full circuit

### *Schematic Editing Commands*

| | |
|---|---|
| (scissors icon) | Delete tool |
| (hand icon) | Move tool |
| (undo icon) | Undo tool |

### *Analysis Commands*

Simulation → Edit Simulation — Choose the simulation type

- DC op pnt — Voltages unchanging (after selecting Simulation → Edit Simulation)
- Transient — Voltages change with time (after selecting Simulation → Edit)
- AC Sweep — Changes input sinusoidal frequency (after selecting Simulation → Edit)

(run icon) — Run the simulation

### *View Results*

| | |
|---|---|
| DC | Hover mouse over schematic for voltage, current, power |
| Transient | Click wire to add to oscilloscope display |
| AC Sweep | Click wire to add to Bode plot |

## LAB PROBLEMS

1. Use LTspice to find the voltage across the 10Ω resistor in the following circuit. (Hint: do not forget to assign a ground node in LTspice!)

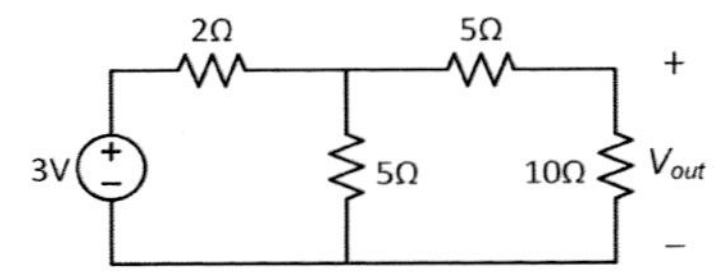


2. The circuit above is powered by 2 AA batteries. How long will the circuit last if a single AA battery holds 10kJ of energy?
3. The **duty cycle** of a square wave is the ratio of the time the cycle is spent "on" divided by the total length of time of each period. For instance, if the "on" time is 1 second, and the "off" time is 3 seconds, then the total period is 4 seconds, the duty cycle is 1/4 = 25%, and the frequency is 1/period = 1/4 Hz. Find the resistors needed to build a 555 circuit that oscillates at 1kHz at a 33.3% duty cycle, then analyze it using LTspice and print out the waveform showing the voltage at the bottom of the LED. Use a 1μF capacitor.
4. Design a second-order Butterworth lowpass filter like Practice Problem 9 with a cutoff frequency of 60Hz using 100k resistors. Analyze it using an AC sweep analysis from 10Hz to 1kHz. Print both the LTspice schematic and the plot of just the output magnitude, not phase, using the Snipping Tool on Windows or Command+Shift+4 on a Mac. You may want use the net method to avoid crossed wires. Hint: if LTspice returns an "unknown subcircuit" error, check to make sure "Universal Opamp2" was used and not "opamp2".
5. Use the five step method of David Jonassen with the NSPE Code of Ethics to analyze the following, unfortunately realistic, situation an engineer may encounter:

   You are an electrical engineer working for a defense contractor. Your team is completing work on the guidance control system for an anti-aircraft missile defense system. Early testing showed that under certain situations the guidance control could malfunction, resulting in unstable missile flight over friendly airspace. The guidance control system was modified to maintain controlled flight in that particular situation.

Although the system now passes all tests, you believe given the earlier failures may point to a deeper problem with the guidance system, that given the right test conditions could again lead to loss of missile flight control. However, you cannot be sure of this; all you know is that the system was redesigned to successfully pass the limited number of flight scenarios proposed at the project outset. When you present your concerns to the company management, they assure you that all tests show that the missile is safe and that further testing would not only be unnecessary but would lead to extremely costly penalties for failing to produce the device on the contract deadline, and may jeopardize the division's (and therefore your job's) existence. They hint that if you fail to approve the system, you may be replaced with a coworker who would be more willing to approve the project.

You feel torn between your duties to your coworkers, your employer, your family, and the military personnel who will be relying on this technology. You do not know if your concerns are valid without months of additional simulations. What do you do?

# INDEX

**N**



**O**



**P**



**R**



**S**